S0-GBG-952

Computer Networking and Chemistry

Peter Lykos, EDITOR
Illinois Institute of Technology

A symposium sponsored by the Division of Computers in Chemistry at the 170th Meeting of the American Chemical Society, Chicago, Ill., Aug. 26-27, 1975.

ACS SYMPOSIUM SERIES **19**

AMERICAN CHEMICAL SOCIETY
WASHINGTON, D. C. 1975

Library of Congress CIP Data

Computer networking and chemistry.

(ACS symposium series; 19 ISSN 0097-6156)
Includes bibliographical references and index.
1. Chemistry—Data processing—Congresses. 2. Communication in chemistry—Congresses. 3. Computer networks—Congresses.
I. Lykos, Peter, 1927-   II. American Chemical Society. Division of Computers in Chemistry. III. Series: American Chemical Society. ACS symposium series; 19.

QD39.3.E46C64   542'.8   75-35538
ISBN 8412-0301-6   ACSMC8   19 1-237 (1975)

Copyright © 1975

American Chemical Society

All Rights Reserved

PRINTED IN THE UNITED STATES OF AMERICA

ACS Symposium Series

Robert F. Gould, *Series Editor*

FOREWORD

The ACS SYMPOSIUM SERIES was founded in 1974 to provide a medium for publishing symposia quickly in book form. The format of the SERIES parallels that of the continuing ADVANCES IN CHEMISTRY SERIES except that in order to save time the papers are not typeset but are reproduced as they are submitted by the authors in camera-ready form. As a further means of saving time, the papers are not edited or reviewed except by the symposium chairman, who becomes editor of the book. Papers published in the ACS SYMPOSIUM SERIES are original contributions not published elsewhere in whole or major part and include reports of research as well as reviews since symposia may embrace both types of presentation.

CONTENTS

Preface vii

1. CRYSNET: A Network for Crystallographic Computing 1
   T. F. Koetzle, L. C. Andrews, F. C. Bernstein, and
   H. J. Bernstein

2. Remote Terminal Computer Graphics 9
   David L. Beveridge and Elias Guth

3. Multiprocessor Molecular Mechanics 17
   Kent R. Wilson

4. Geologic Applications of Network Conferencing: Current
   Experiments with the FORUM System 53
   Jacques Vallee and Gerald Askevold

5. A Network of Real-Time Mini Computers 67
   William J. Lennon

6. A Computer Utility for Interactive Instrument Control 85
   Paul Day

7. Hierarchical Minicomputer Support as a Methodological Aid to
   the Laboratory Investigator 108
   R. L. Ashenhurst

8. Computer Networking at UMR 118
   D. W. Beistel, R. A. Mollenkamp, H. J. Pottinger, J. S. deGood,
   and J. H. Tracey

9. Computer Assembled Testing in a Large Network: The
   SOCRATES System 129
   William V. Willis and J. Seely Jr.

10. The Impact of a Computer Network on College Chemistry
    Departments—The Iowa Regional Network 142
    Warren T. Zemke

11. A Case History in Computer Resource Sharing: *Ab Initio*
    Calculations via a Remote Terminal 153
    D. G. Hopper, P. J. Fortune, A. C. Wahl, and T. O. Tiernan

12. Computer Identification and Interpretation of Unknown Mass
    Spectra Utilizing a Computer Network System 183
    R. Venkataraghavan, Gail M. Pesyna, and F. W. McLafferty

13. Networking and a Collaborative Research Community: A Case Study Using the DENDRAL Programs 192
    Raymond E. Carhart, Suzanne M. Johnson, Dennis H. Smith, Bruce G. Buchanan, R. Geoffrey Dromey, and Joshua Lederberg

14. Networks for Resource Sharing 219
    James C. Emery

Index 231

PREFACE

The term "computer networking" conjures up an image of computer nodes linked by telecommunications in such a way that all those computer resources would be accessible to anyone connected to the net and that a single problem could be handled by two or more computers in the net without human intervention.

Actually the computer networks commonly found today are of two general but simple types: (1) star computer networks, whereby a single, central computer has communication lines radiating to terminal devices; and (2) communication networks, whereby a user at a terminal can connect with any*one* from a variety of computer-based services remotely located, by dialing the appropriate telephone number.

Technically, all the components of computer networking could be present in a single chemical apparatus or laboratory or spread over a campus, a region, a country, or the world.

There are several motivations for computer networking, including:

1. Sharing of data and of bibliographic information resources.

2. Sharing of specialized and expensive computing resources, dedicated to a particular mode of problem solving, by users who are geographically dispersed.

3. Provision for loose-coupled computing support for the experimentalist who has a tight-coupled controlled system and occasionally needs additional computer resources.

4. Standardization of protocols and algorithms. The variations that now exist, which render software from different sources incompatible, are often idiosyncratic rather than substantive.

5. Sharing of algorithms *via* a few sites. This means that the best thinking of researchers working in a specific area can be accumulated, used, and continually revised within a common mode of representation.

6. A distributed network of minicomputers at one site may be cheaper, easier to program, and more reliable than one large computer.

A report has been published on three EDUCOM-conducted broad and intensive seminars on networking entitled "Networks for Research and Education" (Greenberger, M., J. Aronofsky, J. L. McKenny and W. F. Massey, MIT, Cambridge, Mass., 1974) which may be summarized as follows:

1. Computer networking must be acknowledged (no person is an island);

2. major inhibitors of the realization of networking potential are political, organizational, and economic—not technological; and

3. networking by itself is not a solution to current deficiencies—user practices, institutional procedures and government policies need to adapt.

Networking to enhance scientific education and research will develop in direct proportion to the initiative shown by the scientists themselves. Discipline-oriented computing is most likely to promote diffusion and acceptance of computer-based enhancements to the particular discipline. This follows because the hierarchical leadership structure, a common language, the established discipline-based communication media such as journals and conferences—all of which facilitate evolution of a discipline—also provide the natural environment and mechanism for infusing and accepting the comprehensive and pervasive computer impact.

Regional, national, or worldwide networking depends on the quality and accessibility of an adequate telecommunications system. Networking within a laboratory, campus, or city can be implemented directly through telephone lines, coaxial cables, microwave links (line-of-sight, range 50 km) or even infrared beams (line-of-sight, range 10 km, dry and clear atmosphere).

For areas with an inadequate terrestrial communications system, the domestic communications satellite will be important. Because the synchronous orbit satellites will "hover" 22,300 miles above the equator, the total path length of a signal from a terrestrial terminal to a terrestrial computer and back again will be about 90,000 miles. That will require ½ second because of the finite speed of light and will have to be taken into account. Such audio and video communications satellites are already being used experimentally for health care and education in Alaska, covering an area as large as Texas, and for the expansion of medical education into states without a medical school such as Alaska, Montana, and Idaho.

The so-called "march of the minis" is an understatement of the impact of the mini computer—that "shrunk in the wash" *general purpose computer system* now accessible to every university department, every college, and every high school. The mini computer supports not only local interactive computing and batch processing but functions as a remote-job-entry station as well, facilitating access to remote special purpose computing facilities *via* telephone lines—i.e., through networking. Indeed, the hand-held micro computer, as well as "distributed intelligence" microprocessors installed in servo and sensor devices (including all scientific research equipment), puts the choice as well as the computer into the hands of each teacher and researcher; it effectively removes the administrator from computer service selection just as he is removd from the selection of textbooks for a particular course or of a piece of equipment for a research laboratory. So-called "intelligent" terminals give every chemist the opportunity to do local computing and have technologically transparent access to remote computer services *via* telecommunications.

At the other extreme, no university or chemical industry has on its site today the most powerful computer commercially available. Only the taxpayer-supported National Science Laboratories are that privileged. Computer networking means that those expensive resources could be realistically accessible, not only to the in-house scientists physically located at those national laboratories but also to their peers at universities and in industry throughout the nation. The current growing movement toward a National Resource for Computation in Chemistry, if successful, will lead to a truly effective NRCC in direct propotrion to the extent that computer networking is involved.

The papers comprising this Computer Networking and Chemistry Symposium were selected because they span various important ways in which computer networking is affecting chemistry research and education. Although most of the authors are chemists, a few are computer scientists whose work is relevant to chemistry. In some cases the papers are written jointly by chemists and computer scientists, showing what can be done at a well-developed chemistry and computer science interface.

Illinois Institute of Technology PETER LYKOS
Chicago, Ill.
May 1975

CRYSNET: A Network for Crystallographic Computing*

T. F. KOETZLE, L. C. ANDREWS, F. C. BERNSTEIN and H. J. BERNSTEIN
Chemistry Department, Brookhaven National Laboratory, Upton, N. Y. 11973

The CRYSNET system is comprised of a group of five intelligent terminals, which primarily are utilized to carry out computations in the field of crystallography, and which communicate with the Control Data Corporation (CDC) model CYBER 70/76 (7600) computing system at Brookhaven National Laboratory (BNL). These terminals are installed in laboratories geographically distributed from New York to Texas, where they currently are used in a wide variety of structural investigations. Three out of the five terminals have interactive three-dimensional graphical displays. Structural results can easily be visualized through the medium of computer graphics, and the research programs of groups with access to the CRYSNET graphics terminals have been enhanced greatly.

System Organization

The basic CRYSNET system organization has been discussed previously (1,2). Here we briefly will review the hardware and software configurations, with emphasis on some recent developments.
Hardware for the CRYSNET graphics terminals was selected to satisfy two important requirements: 1. The terminal should possess a versatile communications interface and be capable of connection via ordinary telephone lines to a variety of host computers, and 2. The terminal should support a flexible stand-alone interactive graphics package suitable for molecular modeling. The chosen terminal configuration, a Digital Equipment Corporation (DEC) model PDP 11/40 minicomputer, with 28K words core memory, hardware floating-point processor, a 1.2 M word disk, card reader, magnetic tape and printer/plotter, interfaced with a Vector General three-dimensional display, has the additional

---
*Work performed under the auspices of the Energy Research and Development Administration, and supported by the National Science Foundation under Grants AG-370 and GJ-33248X, and in part by the National Institutes of Health under grants CA10925 and RR05539.

advantage that the terminal itself provides sufficient computing power to perform small crystallographic computations.

The graphics terminals currently operate under DOS/BATCH version 9.20c, with applications programs almost entirely written in FORTRAN. Use of a higher-level language enables code to be transferred with relative ease to different hardware configurations, and allows practicing crystallographers to program the system.

The BNL computing center operates the INTERCOM system (SCOPE 3.4) on a CDC 6600 front-end, and SCOPE 2.1 on the 7600. The CRYSNET terminals communicate with INTERCOM <u>via</u> ordinary telephone lines at 2,000 baud (synchronous), under the CDC mode 4c or ANSI 200 User Terminal protocol. INTERCOM provides flexible file-manipulation facilities, allows interactive processing on the 6600's, and submission of jobs to the batch queues of the 6600's and 7600.

Under CRYSNET, large computations may be performed at the BNL CYBER 70/76, by use of programs from an extensive crystallographic library, with I/O especially tailored for a remote user environment. Beside BNL, the CRYSNET terminals have been used to communicate with CDC systems at Courant Institute, New York University, at Lawrence Berkeley Laboratory, and the National Center for Atmospheric Research (Boulder), as well as with the IBM 360/91 at UCLA. An asynchronous port on two of the terminals has been used to communicate with a DEC PDP 10 system at the National Institutes of Health.

<u>Molecular Graphics</u>

A versatile molecular modeling software package is under development for the PDP 11/Vector General terminals (<u>3</u>,<u>4</u>). Currently operational are five major graphics programs:
1. PRJCTN, which allows real-time rotation of stick figures of molecules with up to 740 atoms. This program includes optional side-by-side or top-bottom mirror-image stereo, and provides hardcopy output at a user-designated scale (cm/Ångstrom) on the terminal Versatec printer/plotter. Bond distances, angles and torsion angles of interest may be output to the console keyboard or the printer/plotter. A modified version of PRJCTN, running in interactive mode on the CDC 6600 under INTERCOM, has been used to draw molecules on a Tektronix model 4010 storage-tube graphics terminal. Crystallographic symmetry-generation, currently accomplished via an independent program GEN which produces a PRJCTN input file with symmetry-related atoms, is in the process of being added as an option in PRJCTN.
2. ELLIPS (<u>5</u>), a program which produces representations of molecules with atoms drawn as ellipsoids to show amplitudes of anisotropic thermal motion. These views are similar to those produced by the ORTEP program (<u>6</u>) commonly used on a large computer with a mechanical plotter.

3. STIPL, which displays space-filling models with variable van der Waals radii. Atoms are represented as stippled spheres; the hidden portions of atoms are optionally not drawn. Although the direction of view can be selected by the user, real-time rotation is not possible with this program, because of the relatively large amount of core needed to store the large number of points used to draw most molecules.
4. BUILDR, a model-building program which generates stick figures of molecules from a specified connectivity plus bond distances, angles, and torsion angles. This program is important, because it provides an alternate method for specifying molecular structures, besides the input of a known set of atomic coordinates.
5. MANIPL (7), a program to accomplish the interactive fit of molecular models to contoured electron-density maps. The user can change the contour-level of the map, and control the position, orientation and conformation of the molecule with control-dials. The functions assumed by these dials may be dynamically redefined in order to provide flexibility of operation. Experiments are underway to fit a protein model to a density map with MANIPL (see below), and improvement of this software continues.

Applications

The crystallographic groups who are members of CRYSNET represent a wide range of interests in structural chemistry and biochemistry (Table I). This broad user community requires a comprehensive crystallographic software package to run on the CYBER 70/76 at BNL. Computations recently performed on the CYBER 70/76 have included, for example, phase calculations from multiple isomorphous replacement X-ray data for the protein D-xylose isomerase, full-matrix least-squares refinements of the structures of several carbohydrates (8,9) and an anti-cancer drug (10), and fast-Fourier transform calculations of electron-density syntheses for various hemoglobins. In addition, several structures have been solved with computations performed entirely on the PDP 11/40 terminal. This mode of operation is quite quick and convenient; for example, a complete structure determination for guanidinium trans-disulfatotetraaquocobaltate(II), which included data reduction, and location of all atoms from several successive structure factor-Fourier calculations, was accomplished in this way within the space of one day.

The CRYSNET Vector General displays have been used to portray results at various stages in the process of structure determination and refinement. The ability to obtain graphical output quickly and easily, with optional hard copy, has been especially useful in surveys of trial structure solutions, and in the com position of final results for publication (Figs 1-3). In addition, the display has been used to examine structures from the crystallographic literature, by retrieval of information from the

Table I. Research Interests of CRYSNET Member Groups

Brookhaven National Laboratory
  amino acids and peptides
  carbohydrates
  clathrate hydrates
  electron-density studies
  folate inhibitors
  metal hydrides
  nucleosides

Institute for Cancer Research
  anti-cancer agents
  model compounds for protein-
    nucleic acid interactions
  D-xylose isomerase

Johns Hopkins University
  hemoglobins
  immunoglobulins

Medical Foundation of Buffalo
  antibiotics
  prostaglandins
  steroids

Texas A & M University
bile pigments
coordination compounds
porphyrins
staphylococcal nuclease

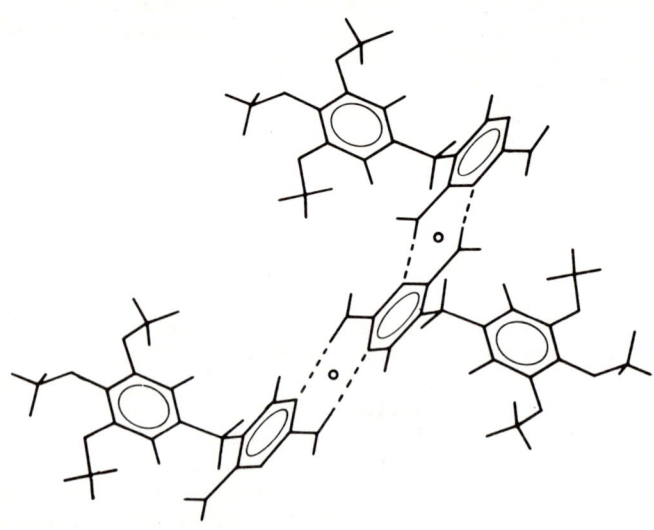

Figure 1. View of the structure of trimethoprim (9); retraced from Versatec plotter output of program PRJCTN

Figure 2. Drawing of L-phenylalanine cation in L-phenylalanine hydrochloride (14); output of program ELLIPS

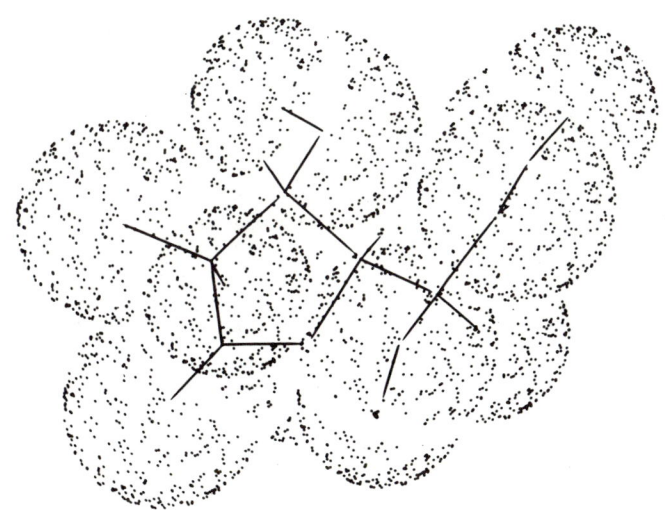

Figure 3. Space-filling model of D-glucaro-3-lactone (8); output of program STIPL

Cambridge Crystal Data File (11) and from the Protein Data Bank (12).

The development of program MANIPL, to accomplish interactively the fit of molecular models to contoured density maps, has been referred to above. To date, segments of a model of staphylococcal nuclease have been fitted with moderate efficiency to a 1.5 Ångstrom resolution electron-density map of the protein (Fig 4). With an entirely different software package, the display has been used to aid the refinement of the structure of lamprey hemoglobin. In this work, models of portions of the polypeptide chain which were idealized by the method of local change (13), have been viewed superimposed upon contoured sections of electron density (Fig 5).

CRYSNET has currently been in operation for approximately two years. During this period, it has become clear that an intelligent graphics terminal can satisfy the computational needs of a typical crystallographic group, and that access to graphics greatly enhances research on crystal and molecular structure, for small and large molecules alike.

Acknowledgement

The authors wish to thank their colleagues H. M. Berman, H. L. Carrell, J. C. Hanson, E. F. Meyer, Jr., C. N. Morimoto, and R. K. Stodola, all of whom have made invaluable contributions to CRYSNET. We also thank M. E. Gress for supplying one of the Figures.

(Abstract)

The CRYSNET system is comprised of a group of five intelligent terminals, which are used primarily for computations in the field of crystallography, and which communicate with the CDC CYBER 70/76 system at Brookhaven. Three out of the five terminals consist of DEC PDP 11/40 minicomputers with 28K words of core memory and interfaced with Vector General three-dimensional displays. An extensive library of interactive molecular-modeling software is being developed for these terminals. Programs are currently available which allow real-time rotation in stereo of stick figures of molecules with up to 740 atoms, display of space-filling models with variable van der Waals radii, generation of stick models from a specified connectivity plus bond distances, angles, and torsion angles, and interactive fit of molecular models to electron-density maps. The terminals provide access to the data files for organic compounds compiled by the Cambridge, England Crystallographic Data Centre, and to the Protein Data Bank, a comprehensive file of crystallographic data for macromolecules, compiled at Brookhaven, which currently includes data for 20 proteins.

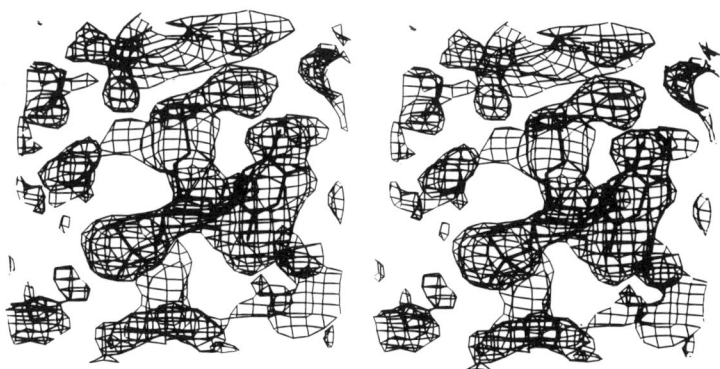

Figure 4. Stereoview of thymidine-3', 5'-diphosphate in single-level 3-D electron-density contours of the ternary complex of thymidine-3',5'-diphosphate, Ca^{2+}, staphylococcal nuclease at 1.5–2.0 Å resolution (map courtesy of F. A. Cotton and E. E. Hazen). Output of program MANIPL.

Figure 5. Stereoview of a fragment difference-density map for lamprey hemoglobin. An adjustment in the position of the proline ring is clearly indicated (map courtesy of J. C. Hanson).

Literature Cited

(1) Bernstein, H. J., Andrews, L. C., Berman, H. M., Bernstein, F. C., Campbell, G. H., Carrell, H. L., Chiang, H. B., Hamilton, W. C., Jones, D. D., Klunk, D., Koetzle, T. F., Meyer, E. F., Morimoto, C. N., Sevian, S. S., Stodola, R. K., Strongson, M. M., and Willoughby, T. V., Second Annual AEC Scientific Computer Information Exchange Meeting, Proceedings of the Technical Program (1974). 148-158. Report BNL 18803, Brookhaven National Laboratory.

(2) Meyer, E. F., Morimoto, C. N., Villarreal, J., Berman, H. M., Carrell, H. L., Stodola, R. K., Koetzle, T. F., Andrews, L. C., Bernstein, F. C., and Bernstein, H. J., Federation Proceedings (1974). 33, 2402-2405.

(3) Andrews, L. C., Pittsburgh Diffraction Conference Programs and Abstracts (1974).

(4) Koetzle, T. F., Andrews, L. C., Berman, H. M., Bernstein, H. J., Carrell, H. L., Meyer, E. F., Morimoto, C. N., Stodola, R. K., and Villarreal, J., 169th National A.C.S. Meeting Abstracts (1975). Abstract COMP 12.

(5) Carrell, H. L. and Stodola, R. K., Pittsburgh Diffraction Conference Programs and Abstracts (1974).

(6) Johnson, C. K. (1965). Report ORNL-3794, Oak Ridge National Laboratory.

(7) Morimoto, C. N. and Meyer, E. F., in Proceedings of the Int'l. Summer School on Crystallographic Computing (1975). F. R. Ahmed, ed. (Munksgaard: Copenhagen, in press).

(8) Poppleton, B. J., Jeffrey, G. A., and Williams, G. J. B. Acta Cryst. (1975)., in press.

(9) Gress, M. E. and Jeffrey, G. A., Amer. Cryst. Assoc. Meeting Abstracts, Charlottesville, Va (1975). Abstract A2.

(10) Williams, G. J. B. and Koetzle, T. F., Amer. Cryst. Assoc. Meeting Abstracts, Charlottesville, Va (1975). Abstract G4.

(11) Allen, F. H., Kennard, O., Motherwell, W. D. S., Town, W. G., and Watson, D. G., J. Chem. Doc. (1973). 13, 119-123.

(12) Protein Data Bank, Acta Cryst. (1973). B29, 1746.

(13) Hermans, J. and McQueen, J. E., Acta Cryst. (1974). A30, 730-739.

(14) Al-Karaghouli, A. R. and Koetzle, T. F., Acta Cryst. (1975)., in press.

2

Remote Terminal Computer Graphics

DAVID L. BEVERIDGE and ELIAS GUTH

Chemistry Department, Hunter College of the City University of New York, 695 Park Avenue, N. Y., N. Y. 10021

The dominant theme in the development of computer technology in the 1970's is teleprocessing. The interface between user and computer has evolved from punched cards and printed output passed across an input/output desk to job creation and retrieval by interactive terminals for individual users and remote batch entry for groups of users. Communication between user and computer is carried out over ordinary telephone lines. Large central processors serving local and regional computer needs on a remote basis are now common, and several national networks, both federaly funded (ARPANET) and commercially funded (CDC-CYBERNET) are operational. More, including a national center for theoretical chemistry calculations, are contemplated.

The scientific community has greatly benefited from these developments. Improved access to digital computer equipment makes possible a vast saving of valuable user time and can significantly increase individual productivity. In chemistry in particular, scientists carrying out theoretical studies of chemical systems based on quantum mechanics, statistical thermodynamics and molecular dynamics or experiments involving large scale data reduction such as x-ray crystallography and magnetic resonance spectroscopy have been direct beneficiaries of developments in computer technology. Quite often, however, the computer output generated by chemical computations is extensive and unwieldy. For interpretation of results one often turns to computer graphics, i.e. incremental plots drawn by a pen under computer control or storage displays wherein a plot is generated under computer control on a storage display cathode ray tube. Several examples of computer graphics used in interpreting various calculations on chemical systems are shown in Figs. 1-5.

The capabilities for accommodating card input and printed output as well as an interactive link with the central processor for console purposes are provided as standard equipment on minicomputer-based intelligent (programmable) remote job entry terminals. At present however, the facility for routinely dealing with computer graphics on a remote basis in a convenient and economic

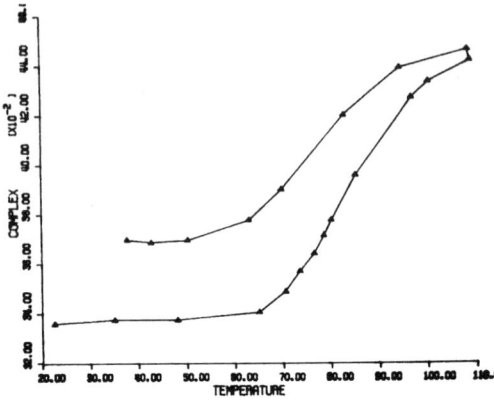

Figure 1. A computer-drawn x-y plot from research conducted by M. Tomasz and co-workers

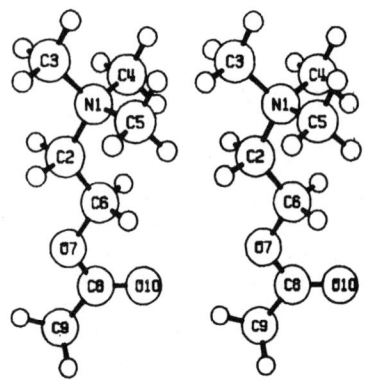

Figure 2. A computer generated stereoscopic view of the chemical neurotransmitter acetylcholine in roughly the geometry preferred in aqueous solution. This drawing was generated using the program ORTEP by C. K. Johnson and was used in studies described by D. L. Beveridge, M. M. Kelly and R. J. Radna, J. Amer. Chem. Soc. (1974) 96, 3769.

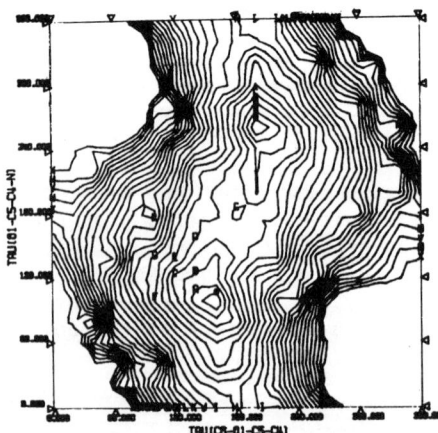

Figure 3. A computer generated conformational energy contour map calculated for acetylcholine in water using theoretical methods. The coordinates $\tau(O1\text{-}C5\text{-}C4\text{-}N)$ and $\tau(C6\text{-}O1\text{-}C5\text{-}C4)$ are dihedral angles referred to Figure 2. The structure in Figure 2 correspond to $\tau(O1\text{-}C5\text{-}C4\text{-}S) = 60°$ and $\tau(C6\text{-}O1\text{-}C5\text{-}C4) = 180°$.

manner is not widely available. We describe herein a method for
remote terminal computer graphics based on a system designed compatible with all standard intelligent job entry terminals, and
completely independent of special consideration from the central
site. The system uses standard commercially available hardware
and can be implemented with a knowledge of minicomputer programming commensurate with that required for laboratory tasks such as
data acquisition and instrument control.

The remote terminal computer graphics system is described as
implemented on facilities of the Hunter Chemistry Computer Laboratory, where graphics capabilities have been developed for network enhanced research and instructional activities based on remote access to the CDC-6600 computer at the Courant Institute of
Mathematical Sciences (CIMS) at New York University and to the
IBM 370/168 facility of the City University of New York (CUNY)
Computer Center. Some background on general aspects of intelligent remote job entry is provided in the next section, followed
in Section II by details of the remote terminal computer graphics
system.

I. Intelligent Remote Job Entry

Remote terminal computer graphics involves additions and
modifications to the remote job entry terminal hardware and software. In this section we describe briefly the hardware and software components of a typical remote job entry terminal. Additional details of teleprocessing for chemists have been described
recently elsewhere.[1]

Hardware. The minimal equipment configuration for intelligent remote job entry is a programmable minicomputer interfaced
to a card reader, line printer, teletypewriter (console), and
data set for telecommunication. A facility of this type can be
implemented for as little as $25,000. A schematic diagram of the
Hunter Chemistry Computer Laboratory hardware is shown in Fig.6.
Note that in addition to the minimal hardware, this system includes a mass storage disc, adding an additional $10,000 to the
capital investment. This is an optional item. The hardware required for remote terminal computer graphics will be described in
the following section.

Software. The computer program resident in the minicomputer
memory during teleprocessing serves three distinct functions:
a) the management of input/output activities, b) the translation
and compression/decompression of outgoing and incoming communications on the data line, and c) data line protocol. These functions are scheduled for parallel processing on a priority interrupt basis. A schematic diagram of typical terminal software designed for teleprocessing in an IBM 370/168 environment[2] is shown
in Figure 7.

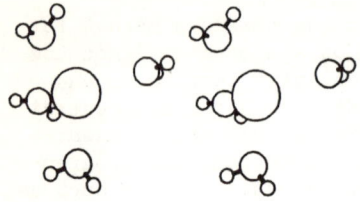

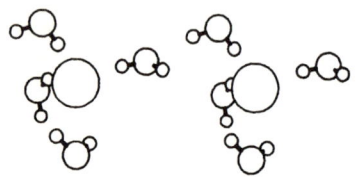

Figure 4. Computer generated stereoscopic views of the minimum energy structure for (top) a tetrahedrally hydrated K^+ ion and (bottom) tetrahedrally hydrated F^- ion. Produced using the ORTEP program in conjunction with studies described by D. L. Beveridge and G. W. Schnuelle, J. Phys. Chem. (1974) 78, 2064 and unpublished data.

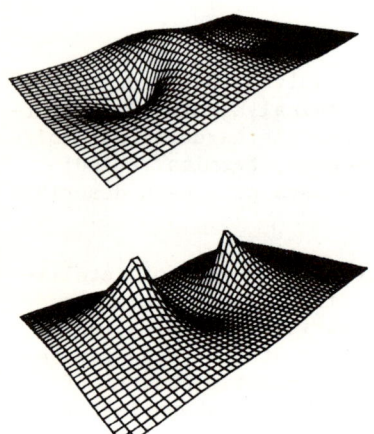

Figure 5. Computer generated perspective views of density difference maps for (top) the σ_g^2 ground state and (bottom) the σ_u^2 excited states of H_2

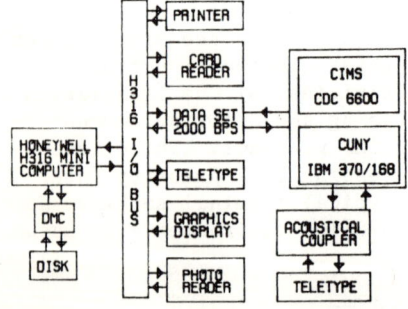

Figure 6. Schematic of hardware elements of the Hunter Chemistry Computer Laboratory ca. May 1975

To follow the flow of events involved in remote batch processing, consider a typical batch job submitted on cards at the terminal and producing printed output. Card input is placed in the card reader and the reader is activated by a console command. Cards are read under program control of the reader driver and copied into the appropriate device buffer. The card image at this point is in a binary representation of hollerith code. When the cards are in and/or the device buffer is full, a flag is set. This schedules the "READER GET" routine which picks up the card image from the device buffer, converts it to EBCDIC, compresses out blanks and repeated characters, and places the translated, compressed code into a communication buffer. When this is complete, the data line transmitter (DLC XMTR), is scheduled. The transmitter section of the teleprocessing software picks up the contents of the communications buffer, adds the appropriate binary synchronous line protocol characters, and outputs the communication character by character to the data line controller and thereby onto the telephone line.

A communication received from the central site is processed under program control of the data line receiver (DLC RCVR) in the teleprocessing software. The receiver picks up incoming communication character by character, interprets and strips away line protocol characters, assembles the reception in a communication buffer, and schedules the "PUT" routine corresponding to the output device for which the reception is intended. The PUT routine manages decompression, translation from EBCDIC to code appropriate for the output device, places the translated, decompressed code in the device buffer, and schedules the appropriate output device driver. The communications cycle is completed when a physical representation of the reception is displayed on the appropriate output device.

For teleprocessing in the environment of another type of host machine such as the CDC 6600 involves an analogous flow of events, but will differ in details such as line code and data line protocol.[3] Thus communication of a single intelligent remote job entry terminal with a multiplicity of different host machines is possible provided the terminal software appropriate for each different environment is available. The facility shown in Fig.6 maintains remote batch entry capabilities with both the CIMS CDC-6600 and the CUNY-370/168 in this manner.

II. Remote Terminal Computer Graphics

The objective of this project was to provide remote terminal graphics capabilities at a level commensurate with standard remote terminal input/output activities as conveniently and economically as possible. Convenience dictates that a) the system should be implemented with no special effort required on the part of the system group at the host facility and the system should operate with a minimum of operator intervention at the terminal

end and no operator intervention at the host facility required.
Economics dictates that the hardware used should be shelf items
and not require elaborate individual modification.

Hardware. The device selected for our system is the
Tektronix 4010 storage tube display terminal, equipped with a
hardcopy device for production of thermofax-type copies. This is
interfaced as an additional peripheral to the terminal minicomputer. The cost of the graphic units is currently around $5000
for the terminal and $3500 for the hard copy device, adding an
additional $9000 to the capital outlay including interfacing.
Thus versatile remote job and graphics capabilities can be obtained with a total capital outlay of the order of $40,000. A
photograph of the remote terminal graphics facility in operation
is shown in Fig.8.

Software. The software involved in remote terminal computer
graphics can be conveniently discussed in terms of a) plot generation and b) plot retrieval and display. The flow of events in
a remote graphics job is shown schematically in Fig.9. This is
typical of our production version implemented in the CUNY IBM
370/168 environment.

To follow the flow of events in remote terminal computer
graphics, consider a typical batch job involving the generation
of graphic output using calls to Calcomp type incremental plot
routines.[4] The job as submitted from the terminal consists of an
application program in compiler language such as FORTRAN and input data. The job is telecommunicated to the central site as described in section II. At the central site the program is compiled. External references to Calcomp plot routines are satisfied using the Calcomp Previewing Routines supported by Tektronix.
The Calcomp Previewing Routines reference in turn the Tektronix
Terminal Control software resident at central site.

Execution of the graphics application program at the central
site causes an input data to be read and generates a standard
printed output file. Calls to the Calcomp subroutines generate
calls to the Terminal Control Program. The Terminal Control Program and attendent display drivers would in a local graphics system manage the creation of a display on the graphics screen,
automatically taking care of sizing and other housekeeping chores.
Here we intervene and require the Terminal Control Program to
produce an output file (plot file) consisting of the arguments to
be supplied to the graphic display drivers for generating the
display. Both the printed output and plot file are then queued
for communication back to the terminal. Any code conversion
scheduled for the plot file by the central site is suppressed. A
high order bit is added to each word of the plot file to distinguish the contents from accidental coincidence with data link
control characters.

In the returning output to the terminal, we are required to

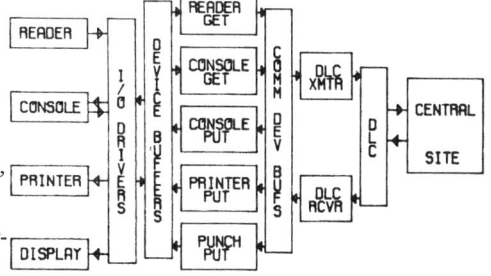

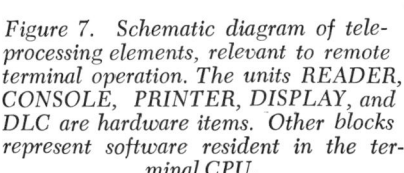

Figure 7. Schematic diagram of teleprocessing elements, relevant to remote terminal operation. The units READER, CONSOLE, PRINTER, DISPLAY, and DLC are hardware items. Other blocks represent software resident in the terminal CPU.

Figure 8. Hunter Chemistry Computer Laboratory remote batch and graphics facility in operation. Users are from left to right Gary W. Schnuelle, Elias Guth and D. Beveridge.

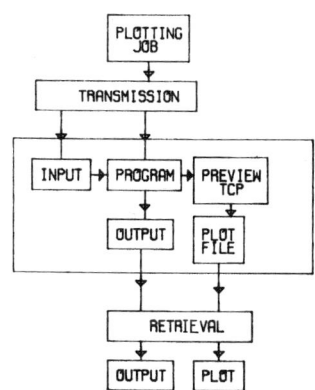

Figure 9. Schematic of flow of events commensurate with a remote terminal computer graphics job

distinguish the plot file from standard printed or console output. The general specification of individual terminal capabilities at the central site may contain provision for any number of peripherals, and customarily IBM 360/20 intelligent terminals and their emulators are support a reader, printer, console and punch. Thus output intended for a certain output device will be preceded by a characteristic data link control character, which is processed upon reception in DLC RCVR. We do not support a card punch locally, so plot files are generated and returned as punch files from the central site. A plot file received at the terminal from the central site passes through DLC RCVR to the plot/punch "PUT" routine where decompression occurs and the file is placed in the graphics terminal device buffer. No code translation is carried out, since the plot file as generated by the Terminal Control Package is already in a form suitable for further use. The graphics display drivers resident in the terminal minicomputer are then scheduled, and the physical display is generated on the storage tube display screen.

The system as described was made operational in the CUNY IBM 370/168 environment in February of 1975. An analogous version has been operational in the CIMS CDC-6600 environment since September 1972. Figs. 1-7 and Fig. 9 of this article were generated on the Hunter Chemistry Computer Facility.

III. Acknowledgements

This project was supported by NSF Grant GJ-32969 from the Office of Computing Activities and in part by a Public Health Research Career Development Award 6K04-GM21281 from the National Institute of General Medical Sciences.

IV. References

1. Beveridge, D.L., Guth, E. and Cole, E.H. in Proc. International Conference on Computers in Chemical Research and Education, D. Hadzi, ed. Union of the Chemical Societies of Yugoslavia (1973).
2. The terminal software for this task was assembled from programs written in part by Mr. Mark Ford of the Honeywell Corporation Automotive Branch of Detroit, and in part locally by the authors.
3. Franceschini, E., Feinroth, Y. and Goldstein, M. AEC Research and Development Report #NYO-1480-148 (1970)
4. "Programming Calcomp Electromechanical Plotters", California Computer Products, Inc. August 1974.

3

Multiprocessor Molecular Mechanics

KENT R. WILSON

Department of Chemistry, University of California, San Diego, La Jolla, Calif. 92037

> "Chemical phenomena must be treated as if they were problems in mechanics." Lothar Meyer, co-discoverer, with Mendeleev, of the periodic table, 1868. (1)

Life is based on marvelous biomolecular machinery, and a central theme in molecular biology has been the role of this machinery's precise molecular architecture. Such molecular architecture, however, is only a static snapshot of a moving, ever changing microscopic world. We are building a macroscopic machine, NEWTON, an internal network of several computer processors, to help the human chemist or molecular biologist investigate and understand the molecular dynamics, the detailed time evolution, as well as the structure of biomolecular systems, for example of enzymes, of membranes, of biomolecular self-assembly. Chemists have long known that such biomolecular behavior is in theory derivable from mechanics, but computational barriers to such a molecular mechanical approach to atomic motions have seemed insurmountable. We have now designed an instrument which will closely interface a human chemist through three-dimensional visual and touch interaction to an extraordinarily powerful networked computer system capable of integrating the classical mechanical coupled differential equations describing the motions of several hundred atoms under their mutual interatomic force fields in such rapid fashion that the chemist can watch the interacting bio- and solvent molecules evolve and reach into a volume of space to actually manipulate simulated atoms, feeling the changing forces upon them, in order to set up and guide the molecular system into the desired chemical pathway. Potential applications span all molecules; their structures, properties and reactions, but particular biomolecular applications include protein conformation, the dynamics of enzyme-substrate interaction, allosteric effects, membrane transport, drug-receptor dynamics and drug design (enzyme blocking agents, antibiotics, specific complexing agents), antigen-antibody interaction and biomolecular self-assembly.

I.   Introduction and Background

Since Newton in the 17th century, it has been clear that the structure (statics) as well as the motions (dynamics) of a set of particles are based on the forces between them. Thus, ever since the atomic theory crystallized in the 19th century, chemists have dreamed that the properties of molecules would someday be derivable from the forces between atoms. We now know that the problem divides into two parts: that a quantum mechanical treatment is required for the electrons in order to derive the potential (or force) field within which the nuclei move; but that for purposes of understanding molecular structure or the mechanism of chemical reaction, once the forces are known the motions of the nuclei may usually be treated by classical mechanics with reasonable accuracy. It is the classical part of the problem, the nuclear motions, which we will discuss here.

While chemists have successfully learned how to understand chemical reactions by calculating atomic motions in a force field for simple reactions involving a handful of atoms, such calculations for reactions of organic molecules of ordinary complexity, let alone the larger molecules of biochemistry, have been too lengthy to handle with present day computer technology. A typical organic reaction including solvent effects might involve 100 atoms, each specified in x, y and z, and thus the solution (in Newtonian or Lagrangian form) of 300 coupled second order differential equations, each with up to 300 variables. Biological polymers, e.g. proteins and nucleic acids, may require thousands of atoms to represent their properties and reactions.

While the amount of computer time necessary to integrate such a set of coupled differential equations is itself a considerable obstacle, the greatest calculational problem which has prevented a "mechanical molecule" approach to organic and biochemical reactions is a more subtle one: the enormous phase space of initial atomic positions and momenta which must be searched to find the region leading to chemically interesting results. Most molecular conformations, relative orientations and relative approach velocities do not lead to the desired chemical reaction. If one were to take a brute force approach and systematically search through just five different initial position vectors and five different initial velocity vectors for a typical 100 atom organic reaction, looking for those which lead to the reaction in question, the set of 300 coupled differential equations would have to be integrated $\sim 25^{100}$ times! Such a brute force approach is not merely difficult with present computers; it is impossible and will remain so.

On the other hand, the well trained human chemist has, or thinks he has, a reasonably good conception of what the atoms must be doing if chemical reaction is to occur. If he could somehow watch the atoms in three dimensions and reach in and adjust or steer the molecules into chemically reasonable proximity, orientation, conformation and velocity, readjusting the calculation in process, he could collapse an impossibly large search space into a manageable size. We have developed a technique of man-machine touch communication to do just this, allowing us to reach into three dimensional space and adjust simulated atoms while feeling the forces involved and simultaneously watching computer generated 3D images of the interacting molecules. A chemist steering the calculation and watching the results will quickly learn, we believe, to converge on the smaller region of chemically meaningful initial conditions. Thus a tight man-machine symbiosis can develop, the human being providing his strong points of recognition of meaningful chemical patterns, of purpose, of direction, of intuition, and the computer providing storage of parameters and rapid and exact calculation of their implications in terms of molecular dynamics, in other words, computing the motions of the atoms in a molecular system under a given set of interatomic forces and steered initial conditions.

If a human is to guide such changing and reacting mechanical molecules, the calculation must be fast enough to match the time scale of human interaction. Even for a molecular system of 100 atoms, only a large, dedicated processor could keep up with a human partner for such a task. To handle several hundred atoms would require a conventional processor of the combined calculational speed of several dozen IBM 360/65's. For larger biological polymers, since the number of instructions scales as N^2, where N is the number of atoms, no present processor appears adequate.

However, as will be discussed below, the needed computational speeds can be obtained with presently achievable logic, memory and bus speeds by distributing the computational load among several fast specialized processors, running in parallel and pipelined modes. Processor costs in instructions/dollar - second have now dropped to the range where biomolecular dynamics can be investigated for molecular systems with a large enough number of atoms to be interesting and important. We have designed such a system, have run a much scaled down test, and are now constructing a first prototype of a full system.

A. Chemistry. The study of the properties of molecules as derived from interatomic forces has developed mainly in two diverse areas: the molecular dynamics of very small molecules in chemical physics and the molecular statics, i.e. the configuration, of larger molecules, particularly the structure of organic molecules and the conformation of proteins.

In chemical physics, one of the major themes of the past decade has been the understanding of chemical processes, including chemical reactions, in terms of the scattering of atoms and small molecules (2, 3). One of the main reasons for the blossoming of this approach to chemistry has been the development of experimental techniques, often involving molecular beams and more recently lasers, capable of studying individual molecular events with sufficient detail and precision that conclusive tests of theoretical scattering predictions are possible.

It is clear that the fundamental theoretical basis of interatomic and intermolecular forces must be analyzed from a quantum mechanical viewpoint (4), as the electrons have small enough mass and momentum that their wave nature cannot be ignored on the molecular scale. It has, however, also become clear in the past decade that if one has in hand an interatomic force function, usually expressed in terms of a multidimensional potential energy surface upon which the more massive nuclei move, a surface which has either been calculated quantum mechanically or arrived at by analysis of experimental measurements, then the calculation of static molecular structure or dynamic molecular evolution and chemical reaction can usually be handled with reasonable accuracy by the approximation of classical mechanics. This type of molecular dynamics approach (5) has been limited to small molecules of only a few atoms because of computational difficulty.

While in chemical physics, molecular dynamics has been derived from interatomic potential (or force) functions, in organic and polymer chemistry molecular structure has been studied on this basis. On the organic molecular scale, researchers have included Hill, Westheimer, Dashevsky, Kitaigorodsky, Allinger, Lifson, Hendrickson, Wibert, Boyd, Schleyer, Mislow and many others (6, 7, 8). This technique has also been extended part of the way into dynamics, to analysis of pathways between structures by calculation of energies at various intermediate configurations in order to study, for example, isomerization (9, 10). In polymer chemistry, particularly protein conformation, a large body of research has been carried out by many groups including those of Ramachandran, Scheraga, Lifson, Flory, Liquori, and Hopfinger (11, 12). Protein conformation is usually treated as an energy minimization problem with constraints. Most bonded interactions between atoms are fixed as constraints in bond length and bond angle, and allowable degrees of freedom are usually rotations about certain bonds. Nonbonded interactions are specified in terms of assumed potential functions. A search is then carried out to try to find the global potential energy minimum, subject to the constraints. This minimum is then assumed to correspond to the most probable conformation for the protein, although more recently the locally available phase space has been estimated as well in order to better taken into account the entropy contribution to the free energy which should more correctly be minimized.

We plan a somewhat different approach, related in coordinates to work by Levitt (13) and by Hermans and McQueen (14), which is easier to generalize to nonpolymers and much simpler to program and to split into tasks for multiple processors. Bonded and nonbonded interatomic interactions will all be treated by the same formalism, not as constraints, but explicitly as force functions, which may however be functions of the vector positions of several neighboring atoms and need not be restricted to two body interactions. As frequently as we can, instead of expressing forces explicitly in terms of angles, they will be expressed in terms of vector operations such as distance and the inner product, which can be calculated much faster than the corresponding trigonometric functions. We will use force functions instead of the equivalent potential functions and will not minimize force or potential energy, but rather assign the correct masses to the atoms and let each one move under the sum of the forces upon it.

It is interesting to further compare energy minimization for molecular statics with molecular dynamics. For large molecules, each type of calculation must spend much of its time in evaluating the potential or force. Thus the additional computer time is not as large as one might think to go from a calculation which searches for a static structure in which the atoms are moved in order to find an energy minimum to a calculation involving a molecular dynamic approach in which actual atomic trajectories are calculated from the accelerations of atoms of defined mass under the forces imparted by other atoms.

Given a machine which will calculate molecular dynamics, one can also calculate structure, by removing energy until atomic motion stops, through the addition of a viscous force proportional to the negative of each atom's velocity vector. This relaxation technique is equivalent conceptually to dunking an initially vibrating molecule into a fluid. As an energy minimum is approached and atomic motion slows down, the viscosity can be decreased to speed approach to equilibrium.

There are two central difficulties well known in energy minimization protein conformation studies which must be dealt with also in using our approach. The first is being trapped in a local (but not global) energy minimum. This is probably less likely than with simple energy minimization schemes, as our atoms will have momentum and are likely to bounce on through many local minima on their way to a deeper minimum. We suspect in addition that the human operator watching the 3D visual display (8, 9) will be able to recognize local trapping as clusters of atoms "catch" on each other, and reach in with the touch interface to nudge them apart. A further serious difficulty is our lack of precise knowledge of the appropriate force functions to use (7, 15, 16). We expect a long and difficult course of research before this imprecision is resolved for organic and biochemical molecules in general and we discuss some directions of approach in a later section. Our machine should be one means of speeding this research.

Classical calculations of the dynamics of large numbers of particles have also been made in at least three other areas: astrophysics (17), plasma physics (18) and the statistical mechanics of assemblies of atoms and molecules. The stellar and plasma mechanics fields share many similarities as well as mutual aspects which differentiate them from the molecular mechanics we are discussing, some of which are summarized in the following table.

Feature	Plasma & Stellar	Molecular
i) Scale of Effects	Often macroscale, gross effects, ∴ many particles needed, less detail.	Microscale, local effects usually dominate, ∴ less particles needed, more precision.
ii) Scale of Interaction	Long range; r^{-1} potentials.	Usually short range; r^{-6} potentials or shielded Coulomb.
iii) Interaction Complexity	2-body usually sufficient; sometimes magnetic as well as electrostatic for plasmas.	2-body sufficient for many interactions, but multibody essential in many cases.
iv) Particle Complexity	Any test particle at given location would feel same force with exception of magnitude and sign (plasma).	Different test particles (elements) would feel quite different force fields.
v) Short-Range Collisions	For many purposes collisions can be ignored.	Collisions essential.
vi) Calculational Techniques	Fourier Transform methods between configuration and field useful.	Fourier Transform methods at least initially appear less useful.

More closely related to our approach are statistical mechanical calculations on large assemblies of atoms or molecules using classical mechanics to follow the trajectories. While most work has used very simplified potentials, such as hard spheres, more recently research has begun with potentials which are quite in the spirit of those needed to consider more complex molecular processes, for example the central force model for water by Lemberg and Stillinger (19) which allows for the possibility of dissociation.

B. Computer. Our design for a hierarchical, distributed, semi-parallel, pipelined computer network follows naturally along the historical path of computer evolution. As

Lorin (20) points out,

> "It is this process of perceiving the existence of smaller and smaller functions and then implementing them in functionally specialized units that is the process of 'maturation' in computer development. The reduction of a view of a system from an undifferentiated mass into a collection of discrete functions is the process by which the computers have matured. Much of this has been made possible by technology but much of it is independent. It is interesting to note there is always a three-stage cycle in design advance: (1) Recognize the function. (2) Identify the structure that could perform it. (3) Build more resources into this structure, i.e. elaborate it, until it evolves into a highly intelligent asynchronous subsystem capable of relieving the 'main path' of considerable burden."

This is the direction we have followed.

As several authors have indicated (21), hierarchical structures, higher levels being composed of similar subunits themselves composed of smaller subunits, etc., are a natural result of the evolution of computer hardware and software, as well as of the evolution of life and of social and economic organizations (including universities). Man himself is a distributed parallel processor of information (22), "a collection of asynchronous subprocessors with highly distributed intelligence throughout the subsystem" (20).

Many parallel processors have been proposed and a considerable variety built (23) for use in pattern recognition, associative processing, optical processing, maximum likelihood calculation, signal processing and the solution of coupled differential equations (24). Our machine, which will be employed for the solution of the coupled differential equations corresponding to Newton's Second Law (and thus the name NEWTON) for a set of interacting atoms, has several antecedents. They include the Lockheed Differential Equation Processor (24), Illiac IV and its Solomon predecessors (23, 24), although Illiac IV seems better suited due to its local data path structure to the solution of differential equations involving fixed neighbors (or intercommunicating cells) rather than to our problem in which the specific neighbors with which each atom must communicate can be continually shifting. A closer resemblance is to the Parallel Element Processing Ensemble (PEPE) being constructed for the Army Advanced Ballistic Missile Defense Agency in a project which has involved Bell Labs, Systems Development Corp., Honeywell and Burroughs (23). It is a highly parallel machine consisting of a general purpose host computer (CDC 7600) and many Processing Elements, each of which can track an incoming missile. It resembles NEWTON in that it requires very fast floating point speeds (~ 1 million instructions per second, MIPS) in each Processing Element and that an important aspect is spatial discrimination, the distinguishing of which objects are nearby. It differs in that the objects are presumed non-interacting. Thus, the structure is single instruction stream, multiple data stream (true parallel) without need for direct element to element communication, while our atoms are interacting and each atom i has its own specific force function, $F_i(r_1 \ldots r_N)$. We must therefore use a multiple (but similar) instruction stream, multiple data stream (semiparallel) distributed structure and provide for intercommunication among the processing elements. In addition, PEPE uses associative memory techniques to distinguish which other objects are the near neighbors of each object, while, for reasons of cost/performance, we plan to use an active discriminator.

Other parallel or distributed systems of interest are the Carnegie-Mellon Multi-Mini Processor (C.mmp), Korn's proposal for a multiple minicomputer differential equation solver, and Neilsen's implementation of such a system. C.mmp is a symmetric set of up to 16 minicomputers (DEC PDP-11's) with an equivalent number of memory units, all interconnected through a cross-bar switch so that any processor can access any memory and up to 16 separate and simultaneous processor memory connections are possible. It is designed to be much more of a general purpose computer than NEWTON, and thus requires hardware and software flexibility beyond our needs. On the other hand, it is not designed for large floating point number crunching, which is our need. Korn (25) in 1972 proposed a system of several PDP 11/45's with floating point processors and shared memories for the on-line solution of coupled differential equations describing dynamical systems. It was designed as a replacement for hybrid digital-analog differential analyzers, and would have been a worthy precursor to the computing engine part of NEWTON if he had been able to build it. While we can run much faster, as will be seen, because of our specialized barrel-roll discriminator and floating point array processors, Korn foresaw many of the aspects of parallel computation applied to coupled differential equation solving which we plan to implement. A system related to Korn's will be implemented at Loma Linda University by Neilsen in the near future and applied to coupled differential equation systems (26).

C. Visual Interface. Already in the 1950's, in realtime military command and control systems, CRT display consoles and a light gun were developed (27). In the 1960's more sophisticated visual display and interaction systems were designed, for example Sutherland's "Sketchpad" (28). Several groups have developed software and hardware for three dimensional computer visual display, including digital and hybrid graphics systems, a head mounted stereoscopic display which moves with the viewer (29) and several visible surface algorithms and hardware processors (30, 31).

These 3D visual output systems stimulated the development of input systems for communicating 3D position to computers, including the Lincoln Wand system using an ultrasonic signal and 4 point microphones (32), the 3D sonic pen and strip microphone system used by Wipke at Princeton to interact with molecular imagery (33), systems employing rotational and translational stages with radial or linear potentiometers (34), and a "Vickers Wand" system using three retractable cords (35), similar in concept to the four cord system we use for the "Touchy Feely", our first touch interface.

Burton and Sutherland developed a system ("Twinkle Box") which allows the real-time measurement of multiple 3D positions (36). Small lights attached to an object (which can for example be a dancing man) flash in sequence, and multiple one dimensional scanners pick up the positions. We plan a related system to pick up both position and orientation with our next touch interface, "Touchy Twisty", designed to allow the user to assemble objects (molecules) while feeling their angular and positional interactions.

D. Touch Interface. In 1965 Sutherland suggested that a computer display system should serve as many senses as possible if it is to provide a "window into the mathematical wonderland constructed in computer memory", and predicted the usefulness of augmenting sight and sound with force display

(37). He suggested implementation of computer controlled kinesthetic presentation through joysticks and mechanical arms.

Two groups, Batter and Brooks at the University of North Carolina and Noll at Bell Labs, first implemented touch communication with computers. Batter and Brooks (38) built and programmed GROPE-1, an on-line interactive computer display system involving two dimensional input to the computer and both visual and force feedback to the user. GROPE-1 was a pilot system having only two degrees of freedom, x and y, and was designed to test the concept of kinesthetic output by exerting programmable forces on the fingers as one moved a knob in a plane, thus enabling the user to examine elementary force fields by experiencing position dependent forces proportional to forces that would be exerted on a particle in a field. The user could both change the position of a particle in conceptual space and feel the force needed to move the particle in the field. In addition to GROPE-1, Batter and Brooks discuss the implementation of kinesthetic display through a mechanical arm, which they have continued to develop.

Noll (34) during the same period built and tested a 3D touch communication system, reporting on it in his thesis in 1971. Orthogonally moving stages are used, culminating in a knob grasped by the hand. Position is picked off by three linear potentiometers and force exerted by three torque motors linking the stages. Software was written in order to simulate objects and surfaces (sphere, cube, sphere within a cube).

Noll discusses many aspects of touch communication, from practical to philosophical, and suggests a variety of applications including studies in perceptual psychology of clash between vision and touch, investigations by motor-skill psychologists, latching onto objects by touch as a means of improving visual displays, testing of manual dexterity, education, design of objects such as telephones with proper "feel" to the hands, aid to the blind, and man-to-man touch communication (for example, a cloth purchaser in New York remotely feeling a manufacturer's cloth in Tokyo).

II. System Design

Our task is to find the set of force functions $\underline{F}_i(\underline{r}_1 \ldots \underline{r}_N)$, $i = 1$ to N which describe the interatomic forces $\underline{F}_i$ as functions of atomic positions $\underline{r}_i$ of N atoms in a biomolecular system and then integrate Newton's Second Law,

$$\frac{d^2 \underset{\sim}{r}_i}{dt^2} = m_i^{-1} \underset{\sim}{F}_i(\underset{\sim}{r}_1 \ldots \underset{\sim}{r}_N), \quad i = 1 \text{ to } N$$

to give the trajectories of the atoms with additional initial steering forces added by the user through the touch interface. Carrying out the task can be divided into two parts: finding the force functions and integrating the equations, chemistry and computation.

A. Chemistry. The most difficult question, one with many clues and a dearth of accurate answers, is, "what are reasonable interatomic force functions which describe interatomic interactions?" Their nature is well described by Lifson and Warshell (39).

"It is commonly assumed that the use of empirical and semi-empirical energy functions in conformational analysis is 'classical,' in contradistinction to quantum mechanical calculations. We suggest that the basic difference is that the latter is a deductive method, seeking to predict observable phenomena from a fundamental law (the Schrodinger equation), while the former is an inductive method, seeking a common analytical representation to a large set of observable phenomena. In fact, there is nothing 'classical' in these functions, as they are not deduced from classical physics. They may just as well be considered as an empirical representation of the Born-Oppenheimer approximation, according to which the ground state of molecules is a continuous function of the atomic coordinates."

The basic assumption which we make, one which indeed is the basis of our general understanding of larger molecules, is that molecules to a reasonable degree of accuracy can be described hierarchically, as assembled from subunits which at least approximately preserve their properties (including force function description) from molecule to molecule. These subunits are usually functional groups or monomers. (This division of molecules into functional groups and monomers is so ingrained in us as chemists that it forms the basis of our molecular nomenclature.) The corollary to this assumption is that the

forces which affect an atom are local, in that there is some
sphere which one can draw around an atom, such that the
positions of atoms outside the sphere have a negligible direct
effect on the central atom. (This is not to say they don't
indirectly affect it by affecting atoms inside the sphere which
in turn affect the central atom.) This localization of interatomic effect is at least implicit in our usual analyses of
molecular structure and only for the case of extended conju-
gated systems or unshielded charges is the sphere likely to be
very large. The implication is that one can find the force on a
particular atom if one knows the positions and natures of the
other atoms within its sphere, independent of the positions of
all the other atoms outside the sphere.

Thus one can hope to develop libraries of subunit force
functions which can provide at least starting points for force
function descriptions of larger molecules.

There are two major paths one can take to try to find
these force functions: calculate them or deduce them from
experimental measurements.

1. Forces - Theoretical. Since the quantum revolution
in the late 1920's and early 1930's, we have known how in
theory to calculate the needed interatomic potential (or force)
functions (4). Despite the immense growth in computer power
in the past two decades, we still cannot practically handle the
full ab initio force function calculations for larger molecules
even on the largest computers. We can, however, handle subunits like the functional groups and simpler monomers to
reasonable accuracy, and this approach has been and certainly
will continue to be a significant source of force data.

In addition, it has more recently become clear that the
forces between non-chemically interacting atoms and groups of
atoms can quite reasonably be derived from localized calculations which require only the wavefunctions of the separate
atoms or groups as input (40, 41). This is an important step
forward, because it reduces a problem which scaled as N^2, in
which N is the number of interacting atoms or groups, to one
scaling as N.

In addition to the ab initio approaches, there are various
semi-empirical theoretical approaches which can be useful.
First, there are semi-empirical methods for solving the quantum mechanical equations themselves. Second, the long range
region of interatomic forces is fairly well understood, for

example Coulomb r^{-2} for charge interaction and Van der Waals r^{-7} for London dispersion forces, and we can calculate such long range forces at least approximately from charge distributions and polarizabilities.

2. Forces - Experimental. Most of our force information, however, will have to come at this stage from postulating reasonable adjustable force functions and then tuning them to match experimental observations of parameters which depend upon these forces. Such methodology has been the basis of much of the conformational calculation work in organic chemistry and polymer chemistry (6, 7, 8, 11, 12, 42, 43).

Many types of physical and chemical measurements can be used to provide force function information and to tune force functions. For example, molecular beam scattering can provide information on both long and short range forces (2, 3) and from transport property measurements one can derive long range forces. Thermochemistry can provide bond strengths and thus well depth information. Many spectroscopic measurements are physically linked to force functions. X-ray diffraction provides structural information and thus the equilibrium geometry where forces balance. Rotational (microwave) spectroscopy can provide accurate equilibrium geometry, bond lengths and angles, and barriers to internal rotation. Vibrational (infra-red and Raman) spectroscopy provides the shapes of force curves near the equilibrium geometry. Electronic spectroscopy (visible and ultra-violet) can provide potential or force function information over a broader internuclear distance range. In fact, any form of spectroscopy, ORD, CD, NMR, etc., in which measurements depend on structure, can be used to tune force functions (44).

Thus, one can hypothesize reasonable force functions for molecules, compare these predictions against physical and chemical measurements and refine the force functions for best match (39). By comparing the results across sets of molecules with the same functional groups or monomers one can build up subunit force function libraries.

We can with NEWTON calculate any interatomic force algorithm, whether given in functional or tabular form, and do it extraordinarily rapidly. For computational efficiency, however, we intend to carefully think about the most efficient way to handle forces for molecular mechanical calculations. It is by no means clear that the usual functional forms for potentials as used in spectroscopy are the most convenient for molecular mechanical calculations.

At least for a start, from the point of view of any given atom, we can divide the other atoms into four categories:

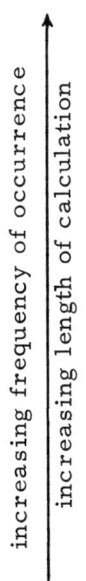

i) those too far away to directly affect it (most other atoms in larger molecules),

ii) those whose effect can be treated by a two body interaction which depends only on the class to which each of the two atoms belongs (for example nonbonded interaction between an H atom on aliphatic C and a ketone O atom),

iii) those which are specifically bonded to the atom in question by bonds which don't change their force function character during the chemical process, and

iv) those atoms whose interaction with the atom in question changes in force function nature as a function of the positions of other atoms, in other words full multibody interactions, for example bond weakening and formation due to the approach of other atoms at active sites in chemical reactions.

Some of the above classification scheme can be reflected in the computer hardware and software in such a way as to vastly increase its calculational speed

B. <u>Computational System.</u> We are now in an era in which if one can understand the structure of a computational problem, one can often design a specialized computational system which distributes the various parts of the calculation among very efficient subunits, such that the computation can be handled much faster and much less expensively than with a general purpose computer.

For example, our molecular dynamics computation has three levels which scale quite differently with N, the number of atoms. The first is the exploration and choice of initial conditions (coordinates and velocities or momenta) of the atoms, which scales as V^N, in which V is the volume of phase space to be explored for each atom. This search space of initial atomic conditions increases as the Nth power, and a brute force approach quickly becomes unmanageable for any computer. The choice of initial conditions is thus much better handled by closely involving a human chemist and relying on his calibrated intuition, his spatial, geometrical sense of what is chemically appropriate. This is the reason for the close attention to visual and touch interfaces: to involve the human chemist in a convenient, comfortable symbiotic relationship with a computer system in such a way as to greatly magnify his ability to solve problems in molecular mechanics.

Second is the discovery of which among the other N-1 atoms are near enough to the atom under consideration to directly contribute to the force on it. This part of the problem scales as N^2, and for larger molecules quickly dominates the computational load. As shown below, we have designed a special barrel-roll discriminator which can solve this part of the computation so rapidly that it no longer represents a major difficulty, at least for the range of N up to a thousand which we are now considering.

Third, given the set of neighboring atoms which are important in determining the force on the ith atom, we must evaluate the force function $F_i(r_1 \ldots r_N)$ for the ith atom and integrate the position r_i of the atom one step forward in time. This calculation for all N atoms scales as N, and is most conveniently carried out on a general purpose computer since for different atoms the F_i can have a variety of algorithmic forms. This part of the calculation we turn over to an expandable set of parallel and pipelined floating point array processors with control, coordination and communication handled by a supervisory computer, as is shown in Fig. 1.

All the parts of this man-machine symbiosis must work smoothly together, to effect the steered solution to the coupled differential equations. It is an initial value problem with man supplying the initial values in real human time and relying on rapid response from the machine to give him feedback so he can trim his input. The machine must rely on man to handle the initial value selection (the Nth power problem) which is beyond machine capability, and the man must rely on the machine to show him the calculated results of his choice rapidly enough so he can use the results to steer the molecules into the desired initial pathway. When the human chemist pushes on certain atoms, the rest of the molecule must follow, and thus the differential equations must be solved rapidly on a human time scale. Our preliminary tests indicate that to match man's visual and touch perceptions, the time to solve one step in the integration should be a maximum of approximately 0.1 second, and this sets minimum performance standards for our machine.

1. Molecular Manipulation - Touch Interface. Touchy Feely I, which inputs 3D position to a computer and outputs 3D force, has been built. Touchy Feely II, which inputs 3D force and outputs 3D position, is being designed, along with Touchy Twisty I which will input three dimensions of position and three of orientation and output three dimensions of force and three of torque. These touch interfaces should be sufficient for most molecular applications.

2. Molecular Portrayal - Visual Interface. The visual interface, the Evans & Sutherland Picture System, is in operation and the software for this application is running. The Picture System allows us to build 3D molecular images, to control atom and bond positions from external computers like NEWTON, to translate and rotate the molecules, to zoom into selected portions, clipping and windowing to see only the chosen region, and to view the molecules if desired in perspective, stereo and color.

3. Distributed Multiprocessor Network. As shown in Fig. 1, the hardware to integrate the coupled differential equations divides into three parts: the supervisory processor, the barrel-roll discriminator and a set of floating point processors.

a) Supervisory Processor. The supervisory processor will be responsible for systems software, for loading and interrogating the barrel-roll discriminator, for compiling and loading microcode for the floating point array processors, for loading their data, for supervising communication among the barrel-roll discriminator, the floating point processors, the visual and touch interfaces and the hierarchical system and its network, and for carrying out or supervising on-line subsidiary calculations based on the trajectories. It needs to be a relatively fast processor, to have sufficient fast and accessible memory to be able to transfer microcode and data rapidly enough to keep the discriminator and array processors busy, and to have enough fixed head disk or equivalent memory to rapidly store trajectory information for later off-line analysis if desired, and a way, perhaps a floppy disk, to conveniently load and store individual programs and data.

b) <u>Discriminator</u>. As shown in Fig. 2, when the number of atoms, N, becomes sizeable, most of the arithmetic operations are taken up by the process of locating the subset of atoms close enough to the atom under consideration to directly contribute to the force it feels. In a dynamic system the identities of the near neighbors can be continually changing. We need only a few out of the several hundred or a thousand atoms for further consideration. The selection task, which increases as $\sim N^2$, soon dominates processing time.

This problem could be handled in several ways: i) by building an appropriate associative memory, ii) by building a special processor, and iii) by being more sophisticated and complex in the programming to take into account "frame coherence", as in the visible surface problem in graphics (30), trying to find ways to use the fact that the integration frames are not independent so that we don't have to do the whole discrimination job over again at each step.

The second alternative, building a special processor to do the discrimination, appears the most cost effective. If we arrange the coordinate data (probably converted to fixed point numbers which we need to do anyway for the Picture System) in the form of a table,

x_1	y_1	z_1
.	.	.
.	.	.
.	.	.
x_N	y_N	z_N

we can then load it into shift registers or incrementally addressed memory and sequentially process it to discover which atoms are within a cube of size $x_i \pm \Delta$, $y_i \pm \Delta$, $z_i \pm \Delta$ centered on atom i at (x_i, y_i, z_i). As a first alternative, we can roll the table into a barrel with x_N following x_1, y_N following y_1 and z_N following z_1 and rotate the barrel by shifting the shift registers around a horizontal axis so that the (x_j, y_j, z_j) pass by a set of parallel comparators which check simultaneously to see if x_j, y_j and z_j fall within the six limits. Several sets of parallel comparators loaded with coordinates of the cubes surrounding different i atoms can all simultaneously be fed the

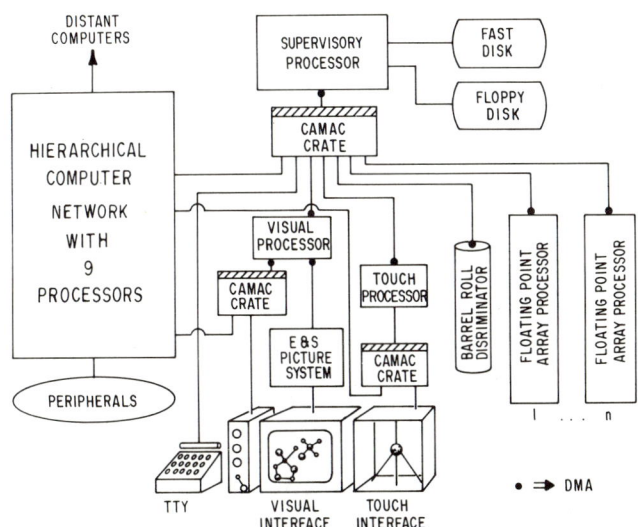

Figure 1. Block diagram of NEWTON. There are six basic subsystems. The Visual and Touch Interfaces provide on-line communication with the user, while the already existing Hierarchical Computer Network, shown in more detail in Figure 5, provides less direct user communication through standard peripherals and also provides network access to other computers on and off campus. The Barrel–Roll Discriminator provides very fast selection of those atoms near enough to each atom to warrant further computation. The Floating Point Array Processors provide very fast (167 nsec pipelined add and multiply) floating point calculation. The Supervisory Processor handles systems software, compiling of floating point array processor microcode, and supervision of other subsystems including their loading and intercommunication as well as error checking. Several of the subsystems already exist: the Hierarchical Computer Network and peripherals, the Evans & Sutherland Picture System, and an initial crude version of the Touch Interface.

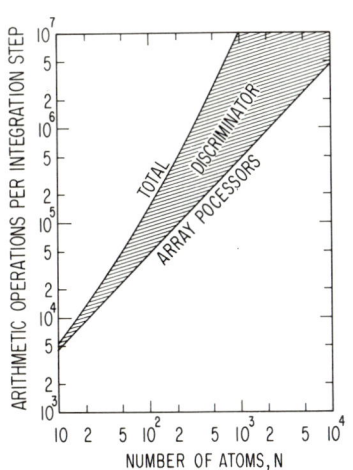

Figure 2. Approximate number of arithmetic operations per integration step as a function of N, the number of atoms. The formula used is: $T = N \cdot [R(N - m) + mP + I]$ in which $T =$ total number of arithmetic operations, $R =$ number of operations for each neighbor rejected by the discriminator, $P =$ average number of operations for each near neighbor through calculation of x, y and z components of force on ith atom, $I =$ number of operations for integration (predictor-corrector), and $m =$ aver. number of near neighbors selected after cubic and spherical discrimination. The figure is drawn using $R = 10$, $P = 70$, $I = 45$ and $m = 6$, and corresponds to force evaluation using an average of 6 near neighbors, each of which takes three functional evaluations by table look-up on higher order bits and linear interpolation on lower order bits. Predictor–corrector integration (Adams–Bashforth) of moderate order is assumed. As can be seen, for N in the range of hundreds or thousands of atoms, most operations can be handled by the discriminator, freeing the more flexible array processors for force evaluation and integration.

table of j atom coordinates as the barrel rolls. A second alternative is to roll the table so that x_1 follows z_1, ..., and x_N follows z_N and rotate the barrel around a vertical axis by N sets of serial adders, each also being fed (x_i, y_i, z_i). In either alternative, all we need out of the discriminator for each atom i is a list of index numbers telling us where in memory to find its near neighbors' coordinates. Either the parallel or the serial version can be constructed from relatively inexpensive components and a relatively modest system can keep ahead of the array processors, even for N's of over a thousand, relieving them, as can be seen in Fig. 2, of the great majority of the arithmetic operations. For a thousand atoms, at 10 integration steps per second, the discriminator must handle $\sim 10^8$ arithmetic operations per second. While this sounds very large at first hearing, the arithmetic task can be shared among many comparators which can run in parallel. In reality the discriminator can be built relatively inexpensively.

c) Array Processors. The evaluation of the force function $F_i(r_1 \ldots r_N)$ given the index numbers identifying the near neighbors for the ith atom and the integration of the acceleration

$$\frac{d^2 r_i}{dt^2} = m_i^{-1} F_i(r_1 \ldots r_N)$$

to give the trajectory for each atom will be carried out by the floating point array processors. In addition, to remove angular bias, a second discrimination will be done, to choose those neighbors falling within the sphere inscribed within the barrel-roll discriminator's cube, i.e. those neighbors for which $r_{ij}^2 < \Delta^2$.

We expect to use commercial floating point array processors, such as those being designed and built by CSPI, by Datawest and by Floating Point Systems. For example, Floating Point Systems' array processor is really an extraordinarily fast, rather general purpose processor, with 38 bit floating point add time of 167 nsec (333 nsec non-pipelined) and multiply time of 167 nsec (500 nsec non-pipelined), and several fast memories. Its 64 bit microcode allows several operations to proceed simultaneously (for example add, multiply, branch). We've coded up some short segments to estimate running times, and its speed is in the class of the largest general purpose processors, yet the array processor fits comfortably on a desk top.

An earlier version is already running plasma particle calculations (45) and use of a network of such array processors is being considered by Lawrence Livermore Laboratory for large scale calculations of the class now being run on CDC 7600 and STAR computers (46).

    d)   Interprocessor Communication. In the evolution toward more and more parallel units, the time spent in interprocessor communication must be controlled, or it can swamp the system. For this reason, one needs a communication network allowing both specific and broadcast modes, so that a processor can access and be accessed by any other specific processor on the one hand and also broadcast information needed by all processors simultaneously, for example the set of coordinates of the atoms at each predictor or corrector substep. We plan to build this dual communication mode capability into the CAMAC system we now use for interprocessor communication, and to increase the speed to match the faster memory speeds.

The above hardware represents an extraordinarily fast computational system for molecular dynamics calculations. NEWTON, as is shown in the following table, can perform these calculations at rates far faster than even a CDC 7600, due to NEWTON's specialized hardware.

ROUGH COMPARISONS OF MILLIONS OF FLOATING POINT INSTRUCTIONS PER SECOND (MIPS)

NEWTON	100 MIPS
CDC 7600	10
IBM 360/195	6
CDC 6400	0.6
IBM 360/65	0.6
DEC PDP 10 (KI-10)	0.6
IBM 370/158	0.4
Interdata 8/32	0.4
Data General Eclipse	0.16
Digital Scientific Meta 4	0.10
DEC PDP 11/70	0.10
DEC LSI 11 microcomputer	0.02

Note: NEWTON can be expanded to several hundred, or even a thousand MIPS by adding more discriminators and floating point array processors, and could then handle a few thousand atoms. NEWTON will probably accomplish those operations to be done in the discriminator by converting coordinates from floating to fixed point.

This is indicative of a new period in computer science, in which specialized computations can be carried out on relatively easy to construct specialized computers at speeds much greater than with general purpose machines.

e) <u>Algorithm</u>. In concept, our task is straightforward. After being presented with the neighbors which matter, the array processor must calculate the force $\underline{F}_i(\underline{r}_1 \ldots \underline{r}_N)$ on the ith atom as a function of its vector position and those of the neighbors. Force function evaluation will normally be by table look-up and interpolation for speed. Then, using Newton's second law,

$$\frac{d^2 \underline{r}_i}{dt^2} = m_i^{-1} \underline{F}_i$$

the acceleration must be used to move forward the trajectory of the ith atom one time step. This must be accomplished for all N atoms within a fixed real time $\Delta \tau$, so that to the human operator the atoms appear to move evenly with time. Predictor-corrector integration methods appear to be the logical choice (<u>47, 48</u>). If only the trajectory in coordinate space is desired, direct second derivative methods which skip the evaluation of the first derivative (velocity) may be used for increased speed. If velocity dependent forces are used for structure calculations by relaxation, or phase space information is desired, then the integration can be performed as two first order equations (Hamiltonian). Discrimination probably need only be performed on the prediction and not the correction substep. The equations are not particularly stiff, as the intrinsic time constants for the bondlength changing modes fall within a reasonably small range. Still, we intend to investigate ways to speed up calculation, either by stiff equation methods or by treating the fast H atom vibrations separately.

While we don't know and probably never will know the basic force functions with great accuracy, we must still proceed in our numerical computation with sufficient precision so as not to lose meaning. Truncation error in the algorithm can be made negligibly small by proper choice of integration order and step size. Error propagation estimates (47) based on presently available floating point array processor word lengths (for example 38 bits, 28 bit mantissa, 10 bit exponent, for floating point systems) indicate that they are long enough for initial work, but that a longer word length will probably ultimately be desirable. One solution would be to calculate the force with single precision and accumulate the integral with double precision (25). Numerous checks can be made on calculational accuracy, including i) change in integration step size, ii) time reversal, iii) conservation laws such as energy, linear momentum and angular momentum and iv) pre- and post-operation by a Galilean transformation, i.e. to and from a frame with a fixed velocity with respect to the original frame (49).

One of the important areas to study will be what practical tradeoffs to make between number of atoms, complexity of force functions, speed and accuracy.

III. Preliminary Test

A crude test of the NEWTON concept has been assembled using available equipment, and is illustrated in Figs. 3-6. Our first touch interface, shown in Fig. 6, is used to move a selected atom in a molecule, and to simultaneously feel the forces which that atom experiences from the other atoms and its own inertia. The motions of the atoms in the time evolving molecule are viewed in 3D on an Evans & Sutherland Picture System, in stereo and color if desired. The differential equations are integrated by a fast micro-programmable mini-computer, a Meta-4, with micro-coded floating point instructions. For 5 atoms interacting by Lennard-Jones 12 - 6 potentials and a non-optimized FORTRAN program using low order Adams-Bashforth predictor-corrector integration, the Meta-4 achieves integration times of 0.05 sec per step or oscillation periods of the fastest atoms in about a second with reasonable accuracy. This integration time is about in the ballpark of the maximum of 0.1 sec per step necessary for reasonable interface to our human visual and touch perceptions of motion. Tests of the viscous damping approach to molecular structure determination show that such a

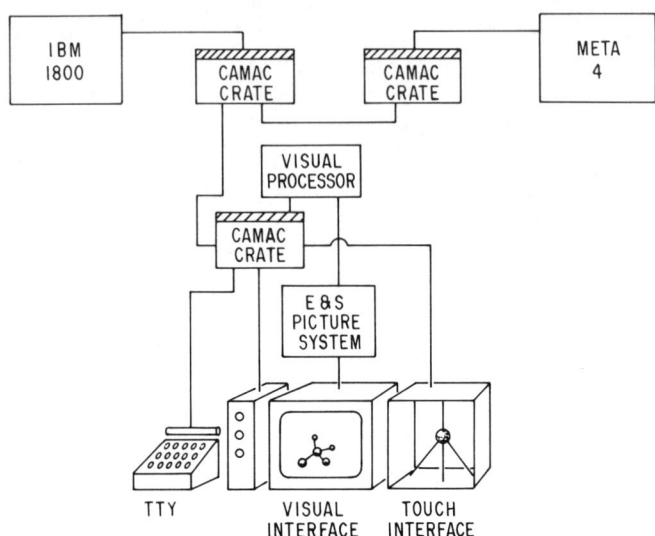

Figure 3. Block diagram of system used to test crudely the concept of NEWTON. The touchstone of the touch interface drives the central carbon atom of a methane molecule, allowing it to be moved and the forces on it from the other atoms to be felt by the user. The molecule is displayed on the Evans & Sutherland Picture System and the differential equations are integrated in real (human) time by the Meta-4 computer to give the trajectories displayed on the Picture System. The Meta-4 is linked through three CAMAC crates and an IBM 1800 to the CDP 135 visual processor emulating a DEC PDP-11/40 which in turn runs symbiotically with the Picture System Processor.

Figure 4. Preliminary test of version of NEWTON using minicomputers, as diagrammed in Figure 3

simple molecule relaxes within a few seconds to its equilibrium geometry. Relaxation is so fast that the structural implications of varying the interatomic force functions can be seen almost as fast as one turns the controlling dial. Since the computation scales as $\sim N^2$, where N is the number of atoms, it can be seen that to handle biomolecular systems with attendant solvent molecules, with hundreds or thousands of atoms, we need a much faster way of carrying out the computation than even a fast minicomputer. NEWTON is designed to provide such speeds.

IV. Applications

A. *Modes of Application.* There are three basic modes of application which we envision for NEWTON: statics, dynamics and statistics.

1. *Statics - Molecular Structure.* As described above, and tested with small molecules, once an intermolecular force field is specified, the atoms moved into roughly their correct positions and a viscous damping applied, they move smoothly and quickly to a minimum energy structure. If the user perceives that this structure is a local and not a global minimum, he can reach in with the touch interface and nudge the atoms along, out of the local minimum.

2. *Dynamics - Chemical Reactions and Evolution of Molecular Systems.* Once the user has specified the intermolecular forces to be used and guided the molecules into the desired initial conditions, he can watch the system evolve in time, reaching in to trim it up en route if he wishes. He can slow it down or freeze frame it at any moment to examine it more thoroughly, and reverse it and step back in time to correct mistakes or repeat a sequence. The initial conditions can be stored and the run repeated or altered, starting with the basis of any previous run. He can turn knobs and zoom in to watch just the dynamics at the active site in more detail. Subsidiary parameters or their functions can be calculated and displayed along with the evolving system, for example the changing force vectors or velocity vectors of atoms as arrows, bond lengths or angles as numbers, energy flow within or between molecules, and progress along user defined reaction coordinates.

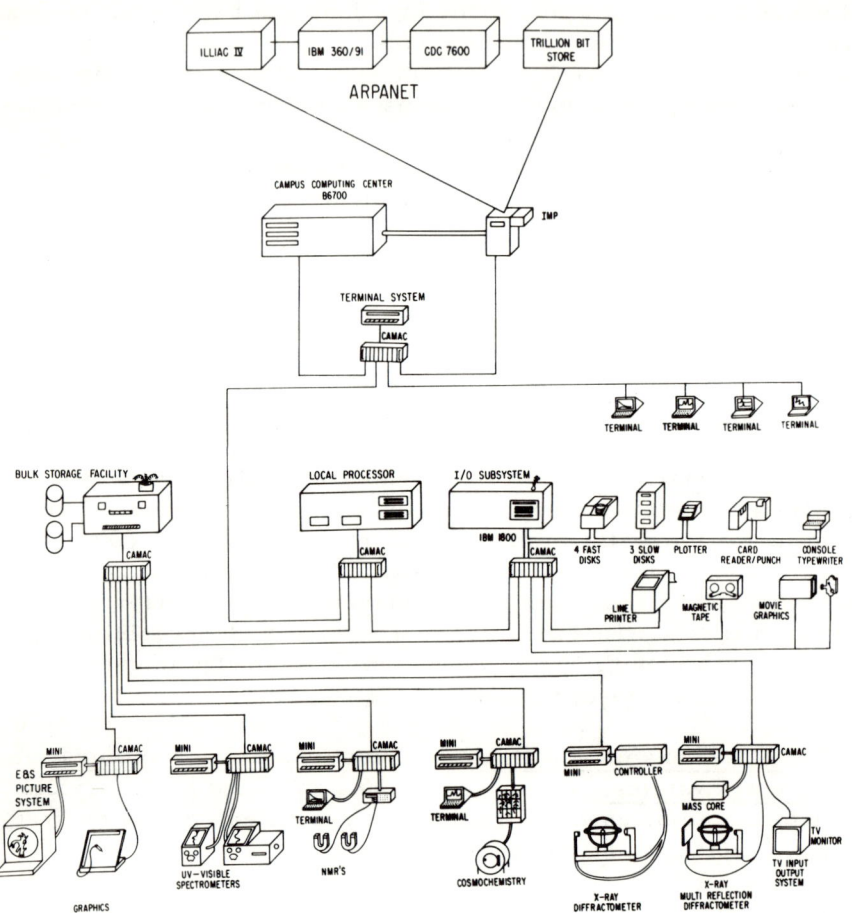

Figure 5. Existing Hierarchical Computer Network used in test. NEWTON will depend upon the Hierarchical Network for large disk support and for most peripherals such as card reader/punch, line printer, tape, and plotter, as well as for links to the campus computer center and off-campus computers and users. The Hierarchical Network links 10 different minicomputers from seven different manufacturers (California Data Processors, DEC, Digital Scientific, IBM, Modcomp, Texas Instruments, and Varian) into a hierarchical three-level system. Level one is composed of various processors running laboratory experiments, and the Evans & Sutherland Picture System. Level two joins together three processors, one providing disk operating systems for all the processors as well as data management out of large disk storage, a second providing local processing on a time shared basis and a third handling the peripherals. Level three is a link to the campus computer center and its Burroughs B6700 and, at least for the moment, a link to the ARPANET. Communication from processor to processor, processor to peripheral and processor to experiment is by a universal hardware system based on the CAMAC convention. This project has involved the design and fabrication of ~ 30 different CAMAC modules, of which ~ 100 units have been built using ~ 10,000 integrated circuits.

3. Statistics - Distributions of Molecular Parameters. In reality, molecules defined by macroscopic conditions do not have one structure but rather a distribution of structures. They do not react by a single reaction path, but rather by a bundle of trajectories. By carrying out sets of runs with NEWTON, we can apply classical statistical mechanics to arrive at such distribution functions. We can invert the usual view of the Ergodic Theorem, and make the connection between measurements we derive over the time evolution of molecules to the distributions to be found in an ensemble of molecules (50, 51). To insure against being trapped in local regions of phase space, we can start different runs from different initial conditions, all corresponding to the same macroscopic set of conditions. Temperature can be set by immersing the molecules in a bath of Ar atoms or H_2O molecules with wrap around boundary conditions and injecting energy, kicking them until the Ar atoms or H_2O molecules (and the molecules of interest in equilibrium with them) reach the desired temperature. In this way entropy, thermodynamic properties and time correlation functions can be studied.

B. Specific Biomolecular Applications. All molecules, their structures, dynamics, reactions, and statistical mechanics, are potential applications for NEWTON and its successors. We will emphasize here the study of the dynamic, as contrasted to the static, function of biomolecules. The version of NEWTON herein described should be able to handle several hundred atoms, and it can be expanded to handle several thousand atoms by adding more of the same distributed processor modules. Many of the applications which are discussed below can be handled by the original NEWTON, while others will have to wait for its more powerful progeny (son of NEWTON = SON or daughter of NEWTON = DONE?), depending upon the number of atoms involved.

Before we discuss particular biomolecular applications, we should clearly focus on the fact that almost all of these molecular systems exist and evolve in aqueous solution and that water molecules and often H^+, OH^- and other ions are importantly involved in what happens. One can try to take into account the solvent effects in various ways (11), but the most appealing to us is to try to directly involve the water molecules and ions by surrounding the biomolecules with a box filled with water molecules and occasional ions; the box having periodic or wrap around boundary conditions so that, for example, each edge of

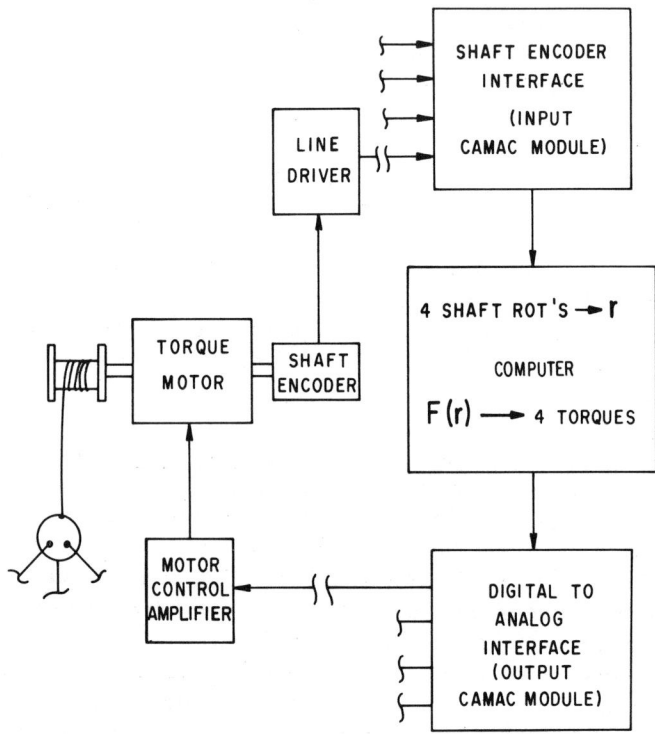

Figure 6. Block diagram of Touchy-Feely I touch interface, with touchstone (central ball) and suspension cables extending to the four corners of a tetrahedron. The position input section measures cable length by means of four cable winches with shaft encoders and reversing counters. The output section impresses force upon the touchstone by driving torque motors on the winch shafts. The computer transforms position and force by matrix operations between ordinary cartesian coordinates and the distorted tetrahedral coordinate system of the touchstone.

the box is also its opposite edge. Explicitly including the solvent will require the ability to process many atoms, but NEWTON can provide this ability.

Some of the many possible biomolecular applications are given in the following list.

1. Protein Structure. A classical area of biomolecular studies is the attempt to understand and predict the relationship between primary protein structure on the one hand and secondary, tertiary and even quaternary structure on the other hand. Using the viscous relaxation damping technique, which is a rapidly convergent process, NEWTON can be applied to the study of these relationships. A practical application of NEWTON would be to the refinement of protein structures derived from x-ray diffraction (13, 52). In addition, conformational transitions and protein folding can be observed and one can study cooperative effects, metastable intermediates, transition velocities and the path to final states (11). Conformational fluctuations can be studied as well as their statistical mechanics (11). Nucleic acid structure could be treated in analogous fashion.

2. Protein Function. As nearly all chemical reactions in living systems are controlled by enzymes, the understanding of the dynamic function of these marvelous machines is an obvious goal for NEWTON users. Allosteric effects may well involve atomic motions which require a dynamic analysis. It is a truism that if one could understand and control enzyme function, one could control life and its diseases. Antigen-antibody specificity could be researched. The self-assembly of proteins and other molecules into more complex structures such as membranes, ribosomes, organelles, and viruses must have a dynamic character, which could be investigated.

3. Membrane Transport. The transport of ions and molecules across membranes is an inherently dynamic process. Ion transport appears to involve electrostatic interactions (53) which are difficult to understand on the basis of plastic, wooden or metal models held in the hand, but can be treated just as easily as steric forces by NEWTON. The acyl chains in the interior of a lipid bilayer membrane can probably largely be modelled on the basis of the well-explored equilibrium force fields for alkyl molecules (8). Perhaps membrane permeation involves not just individual molecular effects, but collective

multimolecular effects as well, in analogy to the collective effects common in stellar and plasma dynamics (17, 18) which would require the modelling of many interacting molecules simultaneously. Models for molecules which act as carriers across membranes (54, 55) can be studied, and series of artificial carriers (56, 57) tried out before the laborious synthesis of the most promising ones. The effects of applied electric fields can be observed, for example in models for the field controlled opening or closing of membrane channels.

    4.   Pharmacology-Toxicology. As mentioned above, control of enzymatic action could be an almost universal tool in disease control. Drug-receptor interaction could be studied in terms of structure and dynamics. Drug design could be aided by NEWTON in the creation of enzyme blocking agents, antibiotics, complexing agents (56, 57) for specific molecules, for example particular metabolites.

Such a broad list of applications, spanning much of biochemistry and molecular biology, is of course more than any investigator can tackle alone. We envision extensive collaboration with other chemists and biologists and in time the replication of more NEWTONS by others, particularly as the cost of later versions can be expected to drop precipitously.

    C.   Other Uses. NEWTON is also applicable to other coupled differential equation problems.

    1.   Chemical Kinetics. Enzyme kinetics and compartment models could be treated, although stiff equation techniques (47, 48) might be necessary.

    2.   Mechanical System Analysis. Prosthetic device design and physiological modelling including blood circulation and organ mechanics studies might be carried out.

    3.   Electrical System Analysis. NEWTON might provide a tool for neural network analysis.

    4.   Fourier Analysis. In addition to coupled differential equation application, NEWTON's floating point array processors will extraordinarly quickly perform Fast Fourier Transforms and thus could be applied to x-ray diffraction analysis for biomolecular structure.

V. Conclusion

A. Networking. We are at the beginning of a new era in computer usage by scientists. Until recently a computer has been something one purchased from a manufacturer, a machine delivered off the shelf and designed to "do all things for all men." Now we are given the opportunity by the advance of computer technology to build our own computer system to fit our own specific needs, much better and less expensively than any general purpose machine. Thus scientists can build special computers for special tasks, just as they traditionally build other special instrumentation to fit their needs. We can purchase subunits on many levels, from integrated circuits to processors, or design our own subunits again ranging from integrated circuits to processors.

As the price of the subunits drop, attention inevitably turns to inter-relationships; how best to assemble subunits into higher levels of hierarchical systems, and in particular, to the intercommunication among subunits. Thus networking of subunits into larger systems naturally comes to occupy our focus as the subunits become more easily available.

This article presents one example of such a networked system. NEWTON, a tight network of processors, some purchased as units and some built from components, will handle a specific molecular dynamics task. It is supported by a loose hierarchical network of processors which provide data management, peripherals and communication for NEWTON and for many other tasks. In turn, the hierarchical network reaches out across the country to access a network of computer installations which can provide a broad menu of additional capabilities.

B. Significance. The relation between structure and function in biomolecules has provided one of the most stimulating areas of science in the past two decades. We may now be at the beginning of an equally exiciting new period, the elucidation of structure-dynamics-function. We believe that instruments such as NEWTON can be important tools for such elucidation, allowing the computation and study of the dynamic (and structural) consequences of interatomic force fields as reflected in the time evolution of biomolecular systems. We foresee the time when we can understand the moving machinery of life: the mechanical molecular motions which are involved in transport across membranes, in muscle contraction, in protein folding, in enzyme catalysis, in allosteric effects and in molecular self-assembly.

Given such understanding, one can then attempt to design particular molecular machines for specific tasks in living systems. If one understands viral self-assembly, perhaps one can build a molecule which will disrupt this assembly. If one understands the action of an enzyme used by bacteria and not by man, perhaps one can build a substrate-substitute which will jam the active site mechanism, blocking its use. If one can understand the difference between the molecular genetic machinery in normal and malignant cells, perhaps one can design a molecular machine to shut down malignant cells. All these are very difficult tasks, perhaps a long way from realization. Yet their significance is great, both intellectually to our understanding of life and practically to our control of disease.

Acknowledgement

Support by the Division of Computer Research of the National Science Foundation and by the National Institutes of Health is gratefully acknowledged. We thank J. Cornelius and the staff of the Chemistry Department Computer Facility for the development of the Hierarchical Computer Network.

Abstract

Chemists have long dreamed that chemical properties could be derived in detail from classical equations of motion based on interatomic forces, but computational difficulties have blocked such calculations for larger molecules. The first difficulty is the large space of initial atomic positions and momenta which must be searched for those leading to chemically interesting results. The second problem is the computational time necessary to solve the coupled differential equations. We plan to collapse this search space with closer man-machine symbiosis based on visual and touch communication of spatial and dynamic chemical information and achieve sufficient computational speed with a network of computers, each handling specialized tasks. The internal computer network will handle this particular calculation several times as fast as a CDC 7600, and will interface with our existing local hierarchical network of computers interconnected through CAMAC modules. Preliminary simulation involving 3D man-machine visual and touch communication and computation on a minicomputer indicates that a practical system can rapidly handle calculations of structure, statistical mechanics and molecular dynamics of reaction for molecules involving several hundred to a thousand atoms.

Literature Cited

1. Dickerson, R. E., Gray, H. B., and Haight, G. P., Jr., "Chemical Principles," W. A. Benjamin, Menlo Park, California, 1974, second edition.
2. Ross, J., ed., "Molecular Beams: Advances in Chemical Physics," Interscience Publishers, New York, 1966.
3. Schlier, C., "Molecular Beams and Reaction Kinetics," Academic Press, New York, 1970.
4. Margenau, H., and Kestner, K. R., "Theory of Intermolecular Forces," Pergamon Press, Oxford, 1969.
5. Levine, R. D., and Bernstein, R. B., "Molecular Reaction Dynamics," Oxford University Press, New York, 1974.
6. Williams, J. D., Stand, P. J., and Schleyer, P. v. R., Ann. Rev. Phys. Chem. (1968) $\underline{19}$, 531.
7. Kitaigorodsky, A. I., "Molecular Crystals and Molecules," Academic Press, New York, 1973.
8. Engles, E. M., Andose, J. D., and Schleyer, P. v. R., J. Amer. Chem. Soc. (1973) $\underline{95}$, 8005.
9. Andose, J. D., and Mislow, K., J. Amer. Chem. Soc. (1974) $\underline{96}$, 2168.
10. Hutchings, M. G., Andose, J. D., and Mislow, K., "Empirical Force Field Calculations of Tetraarylmethanes and -silanes. II. Dynamic Stereochemistry," in press.
11. Hopfinger, A. J., "Conformational Properties of Macromolecules," Academic Press, New York, 1973.
12. Blout, E. R., Bovey, F. A., Goodman, M., and Lotan, N., eds., "Peptides, Polypeptides and Proteins," John Wiley & Sons, New York, 1974.
13. Levitt, M., J. Mol. Biol. (1974) $\underline{82}$, 393.
14. Hermans, J., Jr., and McQueen, J. E., Jr., Acta Cryst. (1974) $\underline{A30}$, 730.
15. Ramachandran, G. N., in Bergmann, E. D., and Pullman, B., eds., "Conformation of Biological Molecules and Polymers," p. 1, Israel Academy of Sciences and Humanities, Jerusalem, 1973.

16. Hagler, A. T., Lifson, S., and Huler, E., in Ref. 12, p. 35.
17. Proceedings of the I. A. U. Colloquium No. 12, Cambridge, England, appearing as Astrophys. Space Sci. (1971) 13, 279-495.
18. Alder, B., Fernbach, S., and Rotenberg, M., eds., Meth. Comp. Phys. (1970) 9.
19. Lemberg, H. L., and Stillinger, F. H., J. Chem. Phys. (1975) 62, 1677.
20. Lorin, H., "Parallelism in Hardware and Software," Prentice-Hall, Inc., Englewood Cliffs, New Jersey, 1972.
21. Pattee, H. H., ed., "Hierarchy Theory," George Braziller, New York, 1973.
22. Arbib, M. A., "The Metaphorical Brain," John Wiley & Sons, New York, 1972.
23. Comptre Corporation, Enslow, Philip H., Jr., ed., "Multiprocessors and Parallel Processing," John Wiley & Sons, New York, 1974.
24. Murtha, J. C., Adv. Computers (1966) 7, 1.
25. Korn, G. A., Simulation (1972) 19, 2.
26. Neilsen, I. R., Loma Linda University, private communication.
27. Sackman, H., "Man-Computer Problem Solving," Auerbach, Princeton, 1970.
28. Sutherland, I. E., Proc. AFIPS, SJCC (1963) 23, 329.
29. Sutherland, I. E., Proc. AFIPS, FJCC (1968) 33, 757.
30. Sutherland, I. E., Sproull, R. F., and Schumacher, R. A., ACM Comput. Surv. (1974) 6, 1.
31. Newman, W. M., and Sproull, R. F., "Principles of Interactive Computer Graphics," McGraw-Hill, New York, 1973.
32. Roberts, L. G., MIT Lincoln Laboratory Report, Lexington, Mass., June 1966.
33. Wipke, W. T., and Whetstone, A., SIGGRAPH Report (1971) 5.
34. Noll, A. M., "Man-Machine Tactile Communication," unpublished Ph. D. thesis, Polytechnic Institute of Brooklyn (1971); J. Soc. Inform. Dis. (1972) 1, (2).
35. Vickers, D. L., "Sorcerer's Apprentice: Head Mounted Display and Wand," unpublished Ph. D. thesis, Dept. of Elec. Engineering, University of Utah (1973).

36. Burton, R. P., "Real-time Measurement of Multiple Three-dimensional Positions," unpublished Ph.D. thesis, Computer Science Division, University of Utah (1973).
37. Sutherland, I. E., Proc. IFIP (1965) $\underline{2}$, 506.
38. Batter, J., and Brooks, F. P., Jr., Inform. Process. (1972) $\underline{71}$, 759.
39. Lifson, S., and Warshel, A., J. Chem. Phys. (1968) $\underline{49}$, 5116.
40. Kim, Y. S., and Gordon, R. G., J. Chem. Phys. (1974) $\underline{60}$, 4323.
41. Kim, Y. S., and Gordon, R. G., J. Chem. Phys. (1974) $\underline{61}$, 1.
42. Bergmann, E. D., and Pullman, B., eds., "Conformation of Biological Molecules and Polymers," Israel Academy of Sciences and Humanities, Jerusalem, 1973.
43. Fraser, R. D. B., and MacRae, T. P., "Conformation in Fibrous Proteins," Academic Press, New York, 1973.
44. Madison, S., in Ref. 12, p. 89.
45. Culler-Harrison, Inc., Santa Barbara, California, private communication.
46. Rudy, T. E., submitted to IEEE Special Issue on Parallel Processing, and private communication.
47. Gear, C. W., "Numerical Initial Value Problems in Ordinary Differential Equations," Prentice-Hall, Inc., Englewood Cliffs, New Jersey, 1971.
48. Gear, C. W., Comm. ACM (1971) $\underline{14}$, 176.
49. Armstrong, T. P., Harding, R. C., Knorr, G., and Montgomery, D., Meth. Comp. Phys. (1970) $\underline{9}$, 29.
50. Tolman, R. C., "The Principles of Statistical Mechanics," Oxford University Press, London, 1938.
51. Rushbrooke, G. S., "Introduction to Statistical Mechanics," Oxford University Press, London, 1949.
52. Levitt, M., and Lifson, S., J. Mol. Biol. (1969) $\underline{46}$, 269.
53. Diamond, J. M., and Wright, E. M., Ann. Rev. Physiol. (1969) $\underline{31}$, 581.
54. Szabo, G., Eisenman, G., Laprade, R., Ciani, S. M., and Krasne, S., in "Membranes, a Series of Advances," Vol. 2, G. Eisenman, ed., p. 179, Dekker, New York, 1972.
55. Eisenman, G., Szabo, G., Ciani, S., McLaughlin, S., and Krasne, S., in "Progress in Surface and Membrane Science," J. F. Danielli, M. D. Rosenberg, and D. A. Cadenhead, eds., p. 139, Academic Press, New York, 1973.

56. Lehn, J.-M., in "Structure and Bonding," J. D. Dunitz, P. Hemmerich, J. A. Ibers, C. K. Jørgensen, J. B. Neilands, D. Reinen and R. J. P. Williams, eds., Springer-Verlag, Berlin, 1973.
57. Cram, D. J., and Cram, J. M., Science (1974) **183**, 803.

4

Geologic Applications of Network Conferencing: Current Experiments with the FORUM System

JACQUES VALLEE
Institute for the Future, 2740 Sand Hill Rd., Menlo Park, Calif. 94025
GERALD ASKEVOLD
U.S. Geological Survey, 345 Middlefield Rd., Menlo Park, Calif. 94025

Computer-based teleconferencing is a mode of communication which enables geographically separated users to jointly manage long-term projects, to organize "instant meetings" without the need for costly transportation, and to exchange documents and review position papers between face-to-face sessions. It thus fulfills a function quite different from that of the telephone, TELEX, electronic mail, or facsimile transmission.
Since 1973, the Institute for the Future and a group within the U.S. Geological Survey have been jointly experimenting with this medium of communication. The experiments have used a family of systems--known as FORUM and PLANET--which the Institute implemented first on the ARPA network, later on a commercial network, and most recently on the Survey's own PDP-10 computer in Denver, Colorado. The experience of the Survey in using computer teleconferencing is typical of what can be anticipated when other scientific communities, such as chemists or physicists, begin to use such media, and thus provides some specific examples illustrating the design and use of these systems.

The Conferencing System

The basic idea of FORUM is to allow unhampered interaction of participants under the guidance of an organizer who defines a topic of discussion, assembles a panel of participants on that topic, and presents the material relevant to the subject. Each participant establishes communication with the computer network via a portable terminal with a standard typewriter keyboard. FORUM is able to convey questions and answers, assemble group opinions, protect anonymous statements, and supply other information to, and within, the group while the organizer monitors the proceedings and intervenes as necessary.
In order to illustrate the nature of the interaction made possible by FORUM, it is appropriate to imagine a hypothetical discussion among a group of experts on the subject of the projected availability of mineral and energy resources in the period

1980-1990. The participants are about 20 in number. Among them are planners, economists, geologists, and petroleum experts. Two are specialists in computerized data bases. In addition, there might be representatives from power and utility companies and the president of a mining corporation. The organizer of the conference has experience in dealing with groups and is familiar with the various techniques which can be brought to bear on the elicitation of forecasts and intuitive judgments in areas of high technology.

This hypothetical conference differs from the usual workshop in that the participants are not meeting face-to-face. Instead, they are geographically separated and use a variety of communication media. Some are sitting around a terminal in a Washington, D.C., office building. A geologist is in the computer room of the Branch of Computations of the U.S. Geological Survey in Denver. One of the economists is in his office at Stanford University. Another one may be sitting in his study at home in New Jersey or in London, for that matter. (These experts are in telephone communication with a central operator who can instantly advise them of the status of the conference, of the progress of work done in subcommittees, or of the reasons for any particular difficulty or delay.) The substantive part of the interaction takes place through entries typed on standard terminals. All of the terminals are connected to the network and are controlled by a computer.

The central problem of implementing such a computer conferencing system clearly reduces to that of identifying, defining, and implementing a range of structures under which the participants are able to share information and enter comments into a common computer-storage file.

The implementation of a system like FORUM raises unusual problems of design: a group of experts or decision-makers typically does not have much knowledge of, or interest in, computer technology per se. There is no opportunity to train them in the use of a text-oriented language before the conference. And it is not feasible to ask them to interface with their peers through information specialists because each participant has a unique awareness of the problem at hand and needs to experience direct contact with his data and with other participants in order to perform at the "cutting edge" of his thinking.

When a group of conferees communicates via FORUM, each participant uses a terminal of the type that can be rented for $150 a month or less. Once the terminal has been logged into the network, the user is presented with a list of discussions which he can attend (just as he would if he were to walk into the lobby of a convention center to review the day's program). Having selected an activity, the conferee is given a short background statement describing the activity. He is then free to observe the ongoing discussion, to review past comments entered into the conference, or to start typing his own remarks. At any point during the discussion, a conferee can send a private message to another participant or make an anonymous entry. All of these communication modes can be entered without the participant's having to learn a single command, thus avoiding a major problem of most interactive systems in existence; namely, that system commands get in the way of the person who types and clutter the transcript with extraneous lines that only have meaning for the machine.

An important facet of FORUM conferences lies in the ease with which the participants have access to services outside of the discussion itself: they can, for instance, submit a prepared statement to the rest of the group or insert parts of the discussion into a personal file. They can also draw responses from a data-base system and enter them into the general discussion. Clearly, the level of interaction thus reached is one not found in face-to-face meetings where experts are cut off from their files and personal notes.

The initial tasks in the FORUM project included an analysis of the available resources and a review of the existing terminal technology in terms of character set, plotting symbols, size of frame, speed of presentation, and interface standards. A decision involving the programming language to be used had to be made early; after exploration of the languages available on the PDP-10 under the TENEX operating system, we reluctantly concluded that assembly language was the only suitable medium to gain access to shared files and to control terminal behavior, both functions being critical to our goal. Additional requirements were speed and low central-processor utilization.

Actual development of the conferencing program proceeded through a series of stages identified as "releases." Release 5 (FORUM-5) was the first version that could conveniently support heavy usage by real-world participants. The code had been modified to make the entire program sharable. Performance measurements showed its central-processor utilization ratio to be excellent (one minute of CPU time for two hours of synchronous discussion per participant). Most command-language features became available to the user within the discussion itself, and use of control characters was practically eliminated. The ability to retrieve and display past entries by date, name, content, and range was made available. Network-wide discussions were conducted routinely and included such topics as the design of advanced teleconferencing systems, the transportation/communication tradeoffs, and initial exchanges of research information with the Communications Study Group in London.

FORUM-6, which was introduced on an experimental basis in August 1974 and was tested until December 1974, features a single, integrated command language, a generalization of the concept of a *conference* to make joint authorship and other management tasks possible, and a scheme for handling private messages in a personal user file rather than as part of the main discussion.

In October 1974, the Institute converted the FORUM program

to a commercial network. This release is tailored especially to the business environment and is known as PLANET-1.

Approach to Evaluation

The Institute's approach to evaluation has been founded on the concept of computer conferencing as a means of communication. The criteria for evaluation of a medium of communication typically involve comparison with other media. And since the medium most familiar to the majority of us is face-to-face communication, there is a tendency for it to become the standard of judgment. One needs to exhibit great care in such comparisons because telecommunications media are not necessarily surrogates for face-to-face patterns. It seems more likely that each medium has *its own inherent characteristics* which should not be expected to mimic face-to-face patterns. At the same time, computer-based systems are too often evaluated and analyzed solely in their own terms. In the case of FORUM, we have sought to relate observations of the medium to an external standard--one which can apply to many media--as much as feasible.

In turning to the literature of group communication, however, we do not readily discover general principles or procedures which are easily adopted as "standard." Certainly the literature of group process is broad and provocative, and the potential for relating group process research to communication research is real, though complicated by many factors.

In designing our research, we have sought to answer two sets of questions:

1. What are the operational characteristics of FORUM as a communications medium? What are the characteristic social patterns of FORUM communication, and how might these be altered?

2. What are the likely social effects of communicating via FORUM on the individual and on the group? How can these social effects be measured? How can FORUM be compared to other media?

To answer the first set of questions, we have devised a method for plotting characteristic social patterns and for analyzing the resulting graphs. A sample of this graph appears in Figure 1.

In the second set of questions, the problem of comparison with other media has led to a search for a general taxonomy--that is, a comprehensive classification system for elements of group communication--which could be employed across media in various group communications situations.

The existing taxonomies of group process are primarily oriented toward communication between two persons (dyadic

4. VALLEE AND ASKEVOLD *Geologic Applications* 57

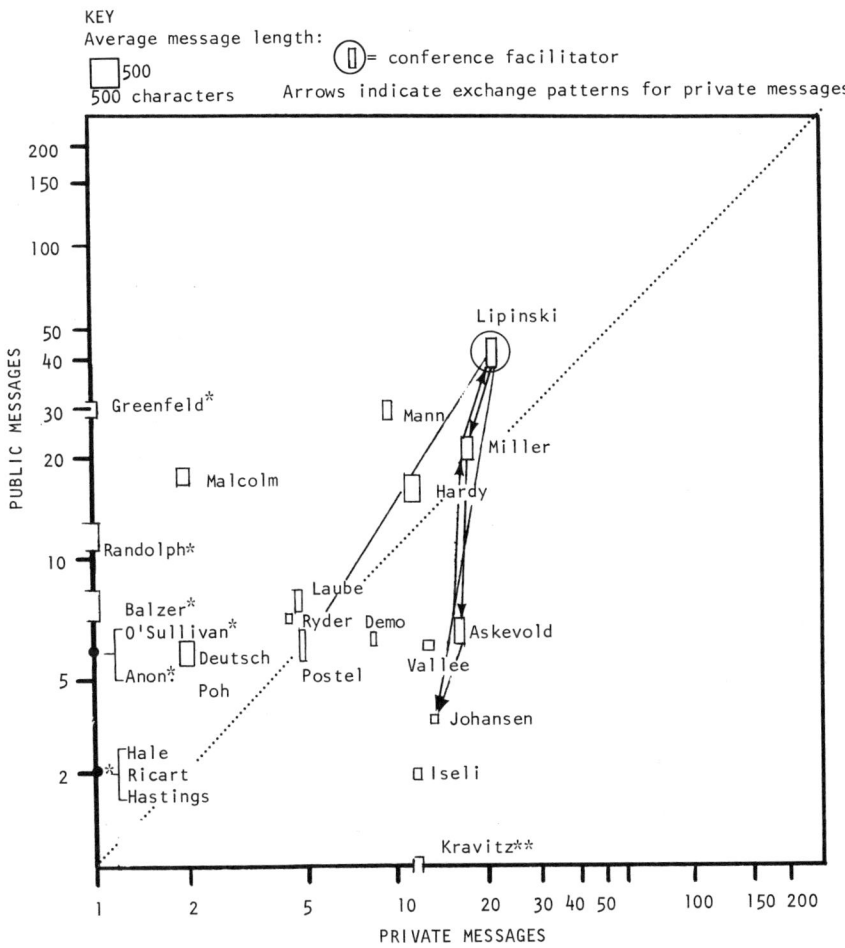

Figure 1. Participation map showing characteristic social patterns

communication). Even though a number of experiments have been categorized as "group" communication, most of these have dealt with the interconnection of two face-to-face groups (i.e., an individual in face-to-face contact with his own group and in contact via electronic media with a single distant group). Extrapolation from dyadic patterns to group patterns, however, is questionable. The principles simply cannot be assumed to be transferable.

In our initial attempt to construct a taxonomy, we have not treated the dynamic aspects of the communication, but have concentrated instead on the elements in a communication situation *before* the interpersonal process begins. Our first, partial taxonomy has thus been arranged to suggest a varied weighting among five key factors--medium, task, rules, person, and group--none of which will be completely discrete. For instance, if members of a given group have a very high need to communicate, they are more likely to make appropriate efforts to gain access to any medium, even if it is difficult to use or unfamiliar to them. Conversely, familiarity with a particular medium is likely to be a very important factor in the choice of that medium for practical communication.

The preliminary results of this social evaluation of computer conferencing are presented in *Group Communication through Computers, Volume 2: A Study of Social Effects*. More definitive results are forthcoming.

Review of Some Early Experiments with the U.S.G.S.

The Survey's interest in creating and using data bases prompted us to begin our experiments by linking mineral resource experts in Washington, Denver, and Menlo Park in teleconferences discussing the present and future availability of fossil fuel and other commodities. Two data-base systems were available online to the participants; one of them was a data base of petroleum reserves which was on the INFONET network, and the other was a catalogue of Alaskan mineral resources stored in the Stanford computer. (Transfer of information in and out of the conference was achieved simply by having two sites operate two terminals, one for the conference and one for the retrieval system.)

The teleconferencing system for these experiments was an early version of the FORUM system on the ARPA network. Although the details of the project have been reported elsewhere ([1]), it is useful to review briefly its conclusions, which encouraged us to enlarge the scope of our joint experimentation:

1. The major advantage of computer conferencing for these applications is the ability to introduce human judgment at a new level in an information system, linking together not only users and sources of data but information experts as well.

2. In computer conferencing situations, group leaders can obtain more deliberate answers to specific technical questions, backed up by facts and with less delay. Both questions and answers are captured on record and can later be reviewed.

3. Computer conferencing appears to be especially useful in coordinating technical projects, when participants are widely disseminated (or traveling extensively) and have a continuing need for reporting and sharing notes.

4. Users of the medium have reported an ability to deal with a larger amount of information more efficiently than through the use of conventional media, such as mail and the telephone.

On the negative side, a major drawback of the early system was the unreliability of the experimental computer network we were using. Access was limited, and frequent hardware failures made "real" work all but impossible. In response to this problem, we initiated two new approaches; namely, reimplementation of an advanced version of FORUM on the Survey's own hardware and research on the feasibility of international conferences using a commercial network.

The FORUM System at the U.S.G.S.: Some Initial Applications

During February and March 1975, a dedicated version of FORUM was mounted on the Survey's own PDP-10 computer in Denver. This installation marked the first instance in which an advanced teleconferencing system had been completely turned over to an operational group.

Figure 2 shows a typical entry process for a FORUM activity. Currently available conferences are presented as a multiple-choice list for the participants. Once a conference has been selected, the agenda is reviewed. In this example, the user (Askevold) was up-to-date in both parts of the discussion. He went to part 1 and requested a listing of the entries made so far. Three entries were found (Figure 3). At this point, another user (Betsy Yount) joined the discussion from her own terminal, and private messages were exchanged (Figure 4). Such messages are not retrievable.

Moving to the second part of the conference, the user again reviewed the transcript from Menlo Park and added an entry requesting that a file of mineral information be loaded into disk storage in Denver (Figure 5). The data base systems under discussion in this conference deal with the chemical analyses of rock samples. An interactive storage and retrieval system named GRASP (developed by Botbol and Bowen of the U.S.G.S.) is used for general geologic applications (Figure 6).

```
RUN FORUM [721,722]

Welcome.

Please type your last name (and then strike the CR key).

- Askevold

Please type your password.

-

Good.  Are you using a terminal that prints on paper?

- Yes

Thank you.

You may attend any one of the following activities:

        1.  Comments on FORUM
        2.  Background on RASS and Test on the Denver System
        3.  FORUM Users Session
        4.  Canada Test of FORUM

Please type the number of the activity you wish to join.

- 2
```

Figure 2. The FORUM System at USGS: Joining a Conference

```
The title of the activity is:
Background on RASS and Test on the Denver System
The parts in the activity are:
    1.  General Information on RASS
    2.  Instructions on How to Access RASS Remotely (Using GRASP)
(To FORUM)
- GO (to part) 1
Part 1
General Information on RASS
You are up to date.
(To FORUM)
- REVIEW (entries) all
[1]  Askevold (Org)  15-Apr-75 9:23 AM
The purpose of this activity is to get some information on RASS which I
don't have, but need prior to my departure to Europe tomorrow.  I will be
sitting in on a meeting of an ad hoc Working Group on Rock Chemical Data
at the UNESCO headquarters in Paris on April 29 and 30 and would like to
contribute something on what RASS is all about, how it is used, and the
role GRASP might have (as Roger and Joe envision it).
[2]  Askevold (Org)  15-Apr-75 9:30 AM
In part 2 I have listed my understanding of how I should arrange to pull
off an Alaska subset of the file as well as how to access what is cur-
rently available for testing; I would appreciate confirmation from some-
one on this along with anything else that might be helpful.
[3]  Askevold (Org)  15-Apr-75 9:32 AM
Joe, Roger mentioned an article on RASS (that you authored); could you
shoot me the reference?

3 entries were found.
```

Figure 3. Reviewing the conference transcript

(To FORUM)
- STATUS (of participants)

Name	Last Time Entered	Last Entry Seen
Askevold	15-Apr-75 3:49 PM	6
Bowen	15-Apr-75 6:25 AM	4
Botbol	Never entered	
Yount	15-Apr-75 3:36 PM	6

(To FORUM)
-
You are now back in the discussion.

(To Yount)
- Betsy, maybe you can give me some of the heaviest users of RASS here in
- Menlo Park for future reference, and some of the problems they run into
- by having to operate in a batch mode, and how going online might solve
- some of these problems.

[7] Yount
Some of the heaviest users of RASS here in Menlo Park are: the people associated with the wilderness programs and the PAMRAP people. The people associated with PAMRAP have no problems, or rather, few problems in using RASS because each quadrangle has a Denver research chemist working with the team leader and they are the RASS interface. Anyone else wanting to use RASS data which they have contributed to--by that I mean the analytical results of their own samples--has difficulty getting the data in.

(To Yount)
-That's great, Betsy, thanks a lot.

Figure 4. Fragments of a synchronous discussion with private and public messages

(To FORUM)
- GO (to part) 2

Part 2
Instructions on How to Access RASS Remotely (Using GRASP)

You are up to date.

(To FORUM)
- Review (entries) all

[1] Askevold (Org) 15-Apr-75 9:33 AM
My instruction to test GRASP on RASS is to: Run IRIS from the Denver machine; how long will this be available? I got into it once, and have to go back to the search examples Roger sent me to complete a valid test.

[2] Askevold (Org) 15-Apr-75 9:36 AM
The data sets for ALASKA are: FIORD.DAT, NEBESN.DAT, and YUKON.DAT. They are protected.

[3] Askevold (Org) 15-Apr-75 9:40 AM
I am to go into SYS F and request MOUNT T658, correct? This should be for a few minutes. Should have two discs mounted. I'm not really sure I have all this right, so please confirm.

3 entries were found.

(To FORUM)
-
You are now back in the discussion.

[4] Askevold (Org)
- Roger, could you please do the above so that this file could be read
- onto the disc space that Mony has set aside for me, and get it off his
- hands? You may have to get in touch with him to get everything
- straight.

Figure 5. Sample transcript with request for a file of mineral information

?
While in the activity you can type an entry at any time. To edit, you may type:

    Control A to delete the last character you typed
    Control W to delete the last word
    Control Q to delete the last line
    Control X to delete the whole entry

In addition,

    Control R will retype the last line as corrected
    Control S will retype the entire message as corrected

To end the entry, strike the carriage return (CR) key twice.

You can send a private message to a participant by typing a left parenthesis "(", followed by his name, a carriage return, and then your message. You can also gain access to special services by sending a private message to FORUM itself or typing a [CTRL] F.

(To FORUM)
- ?
The FORUM services listed below are available to you:

    GO (to part)   FEEDBACK (entries)
    QUIT   JOIN (activity)
    REVIEW (entries)   STATUS (of participants)
    REVIEW (entries)   ADD (participant)
    STATUS (of participants)   REVISE (contents)
    SAVE (entries)   DELETE (entries)
    SUBMIT (file)   ERASE (activity)
    ASK (the following question)

If you do not wish to use any of these services, strike the CR key to return to the discussion.

Figure 6. General user instructions available outline

The ability present in FORUM to elicit online votes and to feedback probability distributions reflecting group judgment in situations involving reserve estimates or exploration decisions represents a new dimension in the use of information systems. These and other user options are readily available to any participant in a FORUM conference.

The system is self-documenting, so that a participant can type a question mark, either during the discussion or in the FORUM services mode, to receive a list of options available at that point. Further documentation is provided when specific services are requested. Figure 6 shows two general overviews of the communication or retrieval options available to the FORUM user.

International Networking

During 1974, we began moving the computer conferencing concept "out of the laboratory" by implementing a conferencing program on a commercial timesharing network. The name of this new program is PLANET, reflecting the major emphasis on joint planning among disseminated user groups. At this writing, the PLANET system has been operational for seven months. It is used by educational institutions in France and in the United States in the coordination of joint computing projects. We have also had experience with several actual "conferences" in which participants made entries through remote terminals over a two- or three-week period instead of traveling to a central location; for such conferences, we have observed a cost reduction of 50 to 60 percent over similar face-to-face conferences.

Since April 1975, we have been holding a continuous computer conference, intended as a computing experiment among members of the COGEODATA community in North America and Western Europe. Figure 7 illustrates the nature of the dialogue in this conference. The reader will note from the time stamps that some entries were made while a user was "alone" in the conference, but others were "synchronous" (entries 20 to 28) with users in Paris and in Menlo Park participating at the same time.

Conclusion

In this paper, we have described two communication systems in current use by a scientific community sharing geochemical and geological information. The first system, named FORUM, is running on the computer of the U.S.G.S. in Denver. The second system, named PLANET, is available to commercial and educational organizations on an international network. Such systems represent a significant tool for the management of joint projects among disseminated groups. They make possible a reduction in travel costs while promoting timely and accurate exchange of data. They also represent an alternative means of publication and a powerful medium for the dissemination of scientific ideas.

[13] Vallee 22-Apr-75 1:39 PM
- Your trip seems to be off to a very good start. You will probably find that telephones are the worst problem in Europe, but we hope to hear from you
- from time to time. Also, Bob Johansen would like to know if you expect to be at the World Future Society in June?
-
[14] Yount 22-Apr-75 3:21 PM
- Gerry, I had the PLANET manual but not the account and keyword information; however, Thad phoned and all is now arranged. I will not be able to phone
- Roger until tomorrow, as it is already 3:30 PM here. I hope that that will be soon enough for your purposes.
-
Things are going smoothly here. One question has arisen regarding connection
- with Stanford. Bruce's feeling is that someone here should write a memo to confirm our interest in getting that service. I will check further with
- Bruce and let you know more about it. If anything needs to be done before you get back, I will write a memo for you.
-
[15] Askevold (Org) 23-Apr-75 9:44 AM
- Thanks a lot, Jacques. Will keep this short. Am using a coupler now rather than a MODEM. Appears to be going okay.
-
[16] Askevold (Org) 23-Apr-75 9:45 AM
- Please apologize to Bob for not responding to him more formally (which I will do with a letter), but if there is a way he can see for me to attend with
- FORUM, but not physically, I would definitely prefer it.

- [17] Askevold (Org) 23-Apr-75 9:47 AM
Thanks, Betsy. We should formally request a line for higher speed work with
- the CRT's. If you can shoot one in, please do so.

- [18] Askevold (Org) 23-Apr-75 9:48 AM
Pleased to report everything went very well--beyond expectations--in London,
- and looks like a good study effort. Will try to come in from Paris...leave

- tomorrow. Many thanks, everyone, and the people here and at ATLAS are very anxious to get in touch with Roger Bowen. Will fill him in as soon as I get
- back.

- [19] Bowen 24-Apr-75 9:43 AM
Gerald--I have the impression from reading your past entries in FORUM (Denver)
- that you didn't get the info in time so I am going to repeat it here just in case.
-
For RASS data, the system has been renamed IRIS. Once you are logged onto
- the system, you access IRIS by: RUN IRIS, etc. (followed by CR)

- There are 5 data sets up. They are as follows:

- (1) RASS1 - An initial data set from Jessie Whitlow.
  (2) ALSKA - All Alaska RASS data (7000 recs)
- (3) YUKON - Data from the Yukon region (>3000 recs)
  (4) NABSN - Data from the Nabesha region (<300 recs)
- (5) FIORD - Data from the Fiord region (<3000 recs)

Figure 7. Sample transcript

- Data sets 1 and 4 are on-line in the public area; data sets 2, 3, and 5 will require a private disk to be mounted. I suggest that (in IRIS) after giving the file command to identify a particular data base, the names command be issued to examine the structure of RASS data sets. Or if you let me know, I'll send a copy of the output you will get using this command.

- [20] Vallee 24-Apr-75 4:18 PM
Gerry, I am preparing the draft of the paper for the American Chemical Society meeting on computer networking and chemistry. I am going over the material we assembled. What about a title like: "Network teleconferencing and Mineral Resource Information: Current Experiments with the FORUM System?"

- [21] Askevold (Org) 28-Apr-75 9:47 AM
I am at this moment in Fontainebleau at the Centre de Recherche Informatique of the Ecole des Mines de Paris. Looking on with me are: M. Kremer, M. Bloch, and M. Lenci. Typical problem now solved: terminals, but usually hard wired. An IBM, but with a French keyboard. Solution: a teletype--voila!

- [22] Askevold (Org) 28-Apr-75 9:49 AM
Actually, the phone noise is not apparent at all. Greetings, Jacques and Thad. See your ESP radar is still on track...a pleasant surprise. What's happening?

- [23] Wilson 28-Apr-75 9:50 AM
Hi Gerry! Summer finally has arrived in Menlo. Is Paris full of the fragrance of spring?

- [24] Vallee 28-Apr-75 9:50 AM
Gerald, I am working on the first draft of the Chemical Society paper. Perhaps I could tell you something about the outline, and you could give me some first-order reactions before I proceed with it. Would you have time for that? As an alternative I can put this into the conference and you can react in the next few days. Which do you prefer?

- [25] Askevold (Org) 28-Apr-75 9:52 AM
I should add that the main purpose of this discussion doesn't have to center around RASS and GRASP....this was only for a special request peripheral to this COGEODATA conference.

- [26] Wilson 28-Apr-75 9:53 AM
Time for me to get back to computing statistics, Gerry. Good talking with you. Will drop in later.

- [27] Askevold (Org) 28-Apr-75 9:55 AM
Jacques, the title sounds fine. Actually I have to catch a bus back to Paris in 15 minutes. I will have access to this same terminal tomorrow, if you can enter what you have and I will review it tomorrow and give you some immediate feedback. Okay? Good talking with you, and will check back tomorrow.

- [28] Vallee 28-Apr-75 9:56 AM
Re 27: Will do. Bonsoir, et salut a l'ecole des mines.

Figure 7. (Continued)

This work has been supported by the U.S. Geological Survey and the National Science Foundation, Division of Computer Research.

Literature Cited

1. Vallee, Jacques, Datamation (1974) 20(5) 85-6, 91-2.

5

A Network of Real-Time Mini Computers

WILLIAM J. LENNON
Computer Sciences Department, Northwestern University, Evanston, Ill. 60201

INTRODUCTION

The principal justification for having a mini or micro computer is the cost effective control and monitoring of real-time systems. A second justification is the relative ease with which independent, dedicated computers can be moulded to fit particular requirements. A common problem which can arise, however, is that when care is taken to improve user interaction or attention is paid to facilitating new program development, expensive additional resources are required.

We solved this problem by building a flexible, resource distribution network to avoid two problems inherent in the routine procedure of expanding the peripheral complement of dedicated computers. First, because the truly expensive resources -- printers, access to the University computer, etc. -- are required only intermittently, these resources will be used inefficiently and generally cannot benefit from economy of scale considerations. Second, and most important, each installation must invest in maintenance and development personnel eventually competing with one another for precious personnel support funds. It has been our experience that we can realize most of the economy of scale inherent in being a computing center without including most of the problems normally associated with use of such an installation. The network has been used routinely since the summer of 1971.

The Computer Science Laboratory at the Technological Institute of Northwestern University is unusual, primarily because it was designed and is maintained, for the convenience of its users. It is a network of mini-computers, which has been designed to provide to remote computers easy access to commonly used peripherals and to the University's CDC 6400 and to support research in distributed resource computing (see figure 1). The continuing growth of the system, and occasional changes that occur, are effected with minimum user inconvenience, as changes to the system do not require concomitant changes in every existing program. Laboratory usage involves both real-time computing and computer system reseach. With few exceptions, software and locally designed hardware are the work of volunteer or student project labor and their contribution to the success of the laboratory cannot be overemphasized. They, in turn, receive experience at a nuts-and-bolts level not attainable elsewhere. The stimulating atmosphere of the laboratory fosters the development of these enthusiasts who spend hours of their own time on projects for faculty members or fellow students. Their volunteer labor, in addition to certain design features of the system, make the network operation exceedingly economical.

Technically, the laboratory houses a star-shaped network of real-time mini-computers, which is interconnected to remote computers, most of which are DEC PDP-8 computers, although there is no restriction as to computer type. The interconnection hardware transceives characters using asynchronous serial communication at a substantial rate (about 14,000 characters per second) with the receiving computer controlling the transmission of each character. This receiver control eliminates the software overhead normally associated with such high speed communication by eliminating the need for complex message handling and character string buffering. Also, utilities and generally needed programs are written to run in the remote, rather than the central computer. The "intelligence" of the network is thus distributed throughout the system, considerably simplifying the maintenance and minimizing the impact of evolving services and facilities.

The advantages of interconnecting a computer to the network are manyfold. The computer gains the ability to perform both qualitatively and quantitatively beyond its own capabilities by using Network Central resources. The cost per computer is very low -- less than that for paper tape equipment, while only under heavy load conditions does its performance degrade to paper tape speeds. Interconnection hardware and software overhead is minimal and the interconnection is easily accomplished. Many of the remote computers have only an instrument interface, a teletype console and a pseudo paper tape connection to the network. In turn, each interconnected computer strengthens Network Central by being available as a distributed intelligence resource to support other computers, when it is not otherwise being used.

The centralized cluster of computers and peripherals is referred to as "Network Central" to distinguish it and the programs required in its operation from the "network" programs, which are generally run in remote computers that are supported by Network Central.

PROCEDURES

Objectives

The goals of the network were to facilitate the design and implementation of new programs as well as generally provide the necessary enhancement to dedicated, real-time computers to make them more flexible, accessible and generally easier to use. These goals were attained by meeting the following four design objectives both for larger machines and minimum configuration control computers which benefit most from access to shared resources.

1. Create an economical system by pooling expensive resources and by providing easy access to it with minimum hardware cost and minimum software overhead.
2. Run all existing programs without modification.
3. Provide a monitor which, while transparent to most programs, would provide full access to all network resources for newly developed programs.
4. Evolve new facilities without interfering with existing facilities.

These objectives have been met for both the minimum configuration control computer, which benefits the most from access to shared resources, and the larger configuration computers.

Network Hardware

The key hardware for the network is a universal serial interface [1] which by suitable logic card or connector changes will transceive 11-unit, 8-bit serial characters at rates between 110 and 153,000 baud into current loops, EIA RS232 interfaces, coaxial cable or twisted pair transmission lines. The hardware uses a reverse channel permission to send signal (see figure 2). Thus, irrespective of baud rate, the receiving device or computer dictates character rate. Time constraints are effectively removed from high-speed communication with an attendant simplification of the required software. The problem of character overruns has been eliminated. The reverse channel is used by each receiver to grant its corresponding transmitter permission to send one character. In our configuration, four twisted pair interconnect computers using differential line driver/receivers.

Commonly used peripherals are installed along with a six million word, moving head disk on a central machine. Time critical or temporary peripherals are installed on two additional central

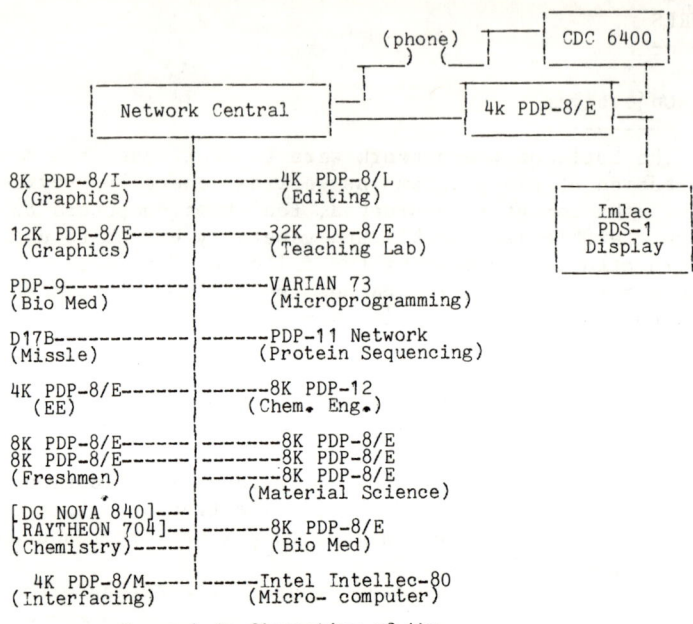

Present Configuration of the
Computer Science Research Network
[Systems scheduled for 1975 expansion are in brackets]

Figure 1.

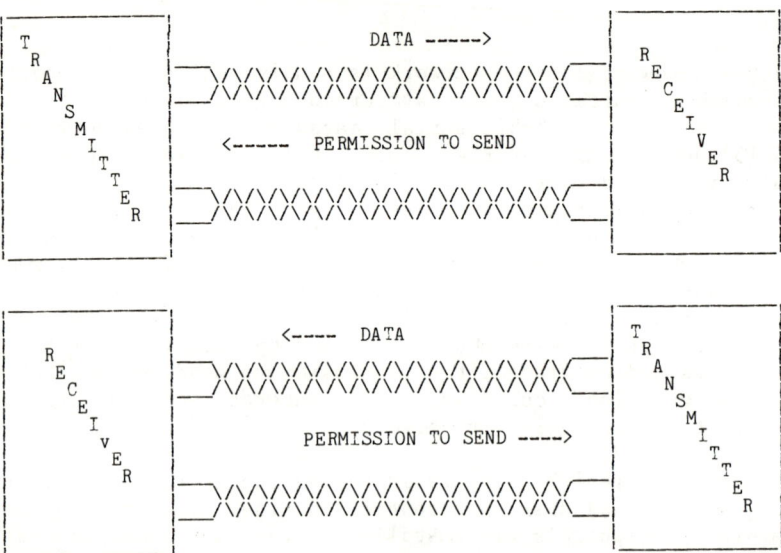

Figure 2.

machines.

Direct access to peripherals is currently denied, although inter-computer communication is allowed for access to specialized equipment. The detailed configuration of Network Central is shown in figure 3. The central computer complex is located in the Technological Institute. It communicates with a dedicated PDP-8/E in an adjacent building which is interfaced to the CDC 6400 and the Imlac PDS-1 computers. The disk is used to store working programs as well as temporary files. A directory records, by file name, storage parameters, as well as creation and usage history. Files are simple linked-list structures of 384, 8-bit or 512, 6-bit character blocks stored as 256, 12-bit words.

Data transmission to remote users is single buffered on a block basis. Character transmissions are executed in a foreground, interrupt handler. Remote communication interrupts are serviced on a round robin basis to provide full-duplex monitor command dialog without degradation during heavy load periods. Buffer completions result in a request to the background support program to initiate disk i/o. Only inter-computer transfers which by-pass the disk approach the hardware design limit of 14K characters a second (i.e. 36 blocks of 384 characters). The maximum rate involving disk transfers is about 7.5 blocks a second. This maximum is realized when only one one-way channel is active (e.g. reading an editor file when no other machines are making demands upon Network Central). While this low density operation is generally the case, transmission rate degrades to about 1 block a second either when a single user runs an input/output bound program or two independent users vie for disk access. Little degradation beyond this first limit is reached by additional users because the disk requests are effectively sorted by position and queued during heavy load periods. The disk track access time is the principal factor in this degradation.

The system design has been purposefully kept simple to permit changes with minimum user inconvenience. Only monitor command processing logic is overlaid on request to facilitate new command feature development. These overlays and the directory are located in the center of the moving head disk. The maximum delay in decoding a request and initiating service currently is two seconds during peak loads. Without any design changes additional users are easily served by additional buffer memory and mass storage. The Network Central program resides in a 12,288 word PDP-8/E in which 5120 words are used as twenty-one disk/user buffers. Each additional 4096 words will provide an additional sixteen buffers. A user computer will require at most two buffers at Network Central. During peak loads when six of the remote machines are performing input/output, CPU utilization approaches 50%. The daily average rarely approaches 10%.

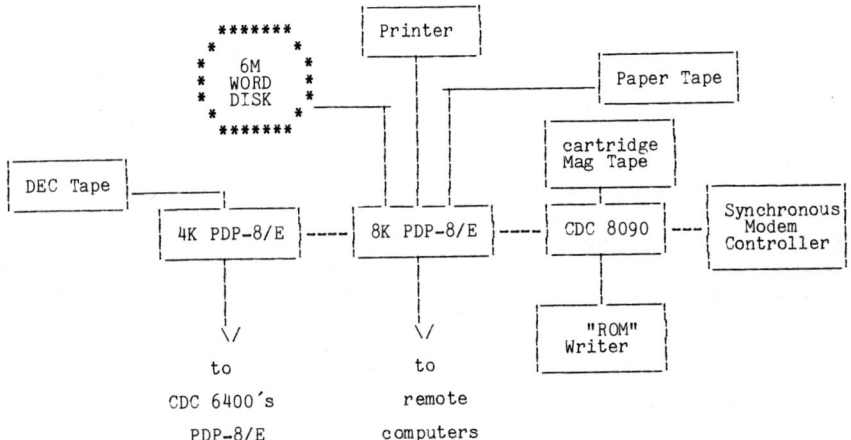

Network Central Configuration
(8090 Peripherals interfaced to simulated PDP-8 external busses)

Figure 3.

Network Programming Characteristics

The principal objective to effectively share resources was easily met. Most paper tape oriented programs can be run without modification. To take advantage of recent manufacturer emphasis on disk operating systems, the command structure of the Network Central monitor includes random access memory operations. A routine complement of driver programs for the "systems device", mass storage, and serial files will generally satisfy the needs of a user seeking to run unmodified mass memory operating system programs -- with or without local mass memory. Simple communication between computers provides access to unique resources and idle computers to allow the easy addition of new support facilities. Finally, the usual complement of utilities and system programs were written to run in remote computers. This facilitates the development of utilities prior to assessing whether a particular function should be included within the complement of Network Central command actions. Transparency, as well as concepts of relative efficiency or effectiveness, are best discussed by examining the individual classes of computer support provided. In this context, transparency is taken to mean the user program is generally unaware of the support provided by the operating system. In particular, only changes explicitly requested are ever made to the remote computer memory holding the user program. Realizing this objective is particularly important when striving to run unmodified vendor programs designed for minimum configuration control computers.

Because most programs written for minimum configuration mini-computers (including similar micro-computers) include paper tape as well as teletype i/o support, network communication paths use control hardware which responds to high-speed paper tape instructions. A unique characteristic of the operating system traditionally used by this class of computer is its minimal resident. While obviously true for a paper tape based control computer, the philosophy is generally retained as mass memory based systems are introduced for the same machine -- programs which ran with only paper tape support are expected to run with the new system. These application programs will often be available only in absolute binary form and cannot be moved to make room for any resident other than one fitting the minimal space originally reserved for the paper tape loader. This philosophy places severe restrictions on any network-connected system designed to support computers running this class of programs.

"Paper Tape" Based Computers

Transparency is achieved by having a resident monitor in the
remote computer occupy the area normally reserved for a paper tape
loader. The monitor resident effectively connects the remote
computer console directly to a Network Central machine where the
majority of the monitor logic resides. Network Central replies are
prefixed and/or followed by non-printing command codes which are
interpreted by the resident while typing the message to the user.
These non-printing characters may be used to automate system
access without resorting to an analysis of the character strings
designed for remote user communication. The resident includes a
modified loader which is controlled by Network Central. After a
program is loaded, a Monitor dialog establishes a source of
characters for the paper tape "reader", as well as a destination
for those "punched." After this dialog, an unmodified program
operates as if it had a 14,000 character per second paper tape
system.

Since the same channel is used for data and control information,
Network Central and locally developed remote programs monitor data
streams for an escape sequence (2 ASCII "CAN" codes) which never
arises in standard "paper-tape" transmissions. The remainder of
the remote programs are usually sensitive to a "reader time-out"
to signal end of file. Files which might accidentally contain the
escape sequence are transmitted in a 6-bit mode. Most of the
Network Central commands are summarized at the end of the paper.
An examination of the commands will confirm the fact that this
class of computers requires the greatest level of support. The
majority of the commands are oriented toward a user communicating
with the Network Central monitor to obtain programs and manipulate
serial files.

"Paper Tape" program examples

In the first example, an unmodified program, the Focal language
interpreter, is loaded and the Network Central files named OUTPUT
and INPUT are assigned to serve the function of paper tape punch
and reader expected by the interpreter. The Network Central
Monitor prompts the user with a carriage return/ line feed and
sharp sign ("#").

        Example 1.

#GET FOCAL   Load FOCAL interpreter.

#ASSIGN OUTPUT<INPUT/G   '/G' is the command to
                             start after assignment.

CONGRATULATIONS!!!   Initial dialog of Focal.

YOU HAVE SUCCESSFULLY etc.

(Typical time to dialog is 15 seconds)

A small mini-monitor (the resident less loader logic) is imbedded in commonly used programs (e.g. Editors, Assemblers, Compilers) and the network drivers of larger, self-supporting operating systems.

The following example shows the monitor and mini-monitor dialog for a a modified "paper tape" assembler used to translate a source program approximately 3000 characters long. The assembler obtains the date and time from Network Central for the listing output. Translation under idle network conditions takes about 40 seconds with the listing pass taking an additional 65 seconds. As should be expected for a program designed for paper tape, the file containing the source is read three times. The second pass produces a binary tape image in the Network Central file "BINARY." The third pass produces a character string stored in the file named "LISTING." When a listing file is diverted to the printer, its name is placed in a queue which must be emptied by an operator at Network Central. In addition, the listing can be viewed using the standard editor program. The "%" symbol is used to distinguish mini-monitor from full monitor dialogs.

                Example 2.

#RUN PALN	Load and start assembler.
%ASSIGN SOURCE	Start pass 1.
%ASSIGN BINARY<SOURCE	Pass 2.
%ASSIGN LISTING<SOURCE	Pass 3.
#DIVERT LPT<LISTING	Queue the listing for print out at # Network Central.

The normal evolution of a Network Central facility begins with a remote computer program that provides service while we obtain usage data, suggestions and an assessment of its usefulness. Few utilities have demonstrated sufficient stability to be included in new versions of the central program. A similar evolution occurs for programs which are modified to use network resources more efficiently. The present assembler has stabilized as the example shows because most users feel that little more would be gained for the effort involved to automate the dialog. Four years experience indicates improvement is due but that it is not a trivial task as the following example shows.

$$\text{Example 3.}$$

#RUN PALN

%ASSIGN SOURCE1

%ASSIGN BINARY<SOURCE1   Remote command.

OUTPUT FILE EXISTS   Error report.

%SCRAP YOUR DIRECTORY   Remote command.

SAME TO YOU FELLA!!   Error retort.

%PURGE BINARY

%ASSIGN BINARY<SOURCE1

etc.

At present, any programmed computer may have the same level of support as a PDP-8 class computer. Adequate disk space must be available to store essential systems and applications programs. A resident, as well as a mini-monitor, must be written for each unique computer. One powerful feature of the network is the growing number of centrally maintained systems and utility programs. To gain access to these older programs, as well as provide access to newer programs for users of the old computers, provision must be made to control programs running on one computer from another. The communications capability is present and as more computers evolve to a size which can support mass memory operating systems having truly device independent input/output, this problem diminishes. Any user at any computer will have the ability to seek out an idle machine to run a program requiring a particular configuration.

We are adding micro computers to the network as if they were inexpensive mini computers. They typically have a network resident and read/write memory. However, as particular applications are defined, there will be a growing number of dedicated processors using read-only memory to support a fixed application -- say, editing. After meeting their principal requirement, the network resident will then be used to communicate with and control other computers running application programs. Successful upgrade of most

network machines to the general capabilities of those supporting
operating systems will mean a general trend away from machine
language to higher level language programming. Newly developed
utility programs will tend to become more machine independent.

Mass Memory Operating System Support

A network connection will simultaneously support either two serial
files or up to sixty-four random files. The random file commands
request the transfer of either subsequent blocks or specific
blocks relative to the beginning of a file. These commands may be
issued by program modules in the remote computer to support a mass
memory operating system.

When considering the support of a mass memory operating system in
the network, three distinct levels are possible. Simplest of all
is supplying a serial file driver to perform pseudo paper tape
support to computers already capable of running the operating
system. Second, mass memory drivers can be supplied which treat
individual files or groups of files as separate mass memory
devices. Finally, a special "systems device" driver can be defined
to use Network Central resources in lieu of any local mass memory.

The details concerning which combination is preferred clearly
depends upon the particular computer configuration and its base
operating system. In the case of the PDP-8, we were able to run
the OS/8 operating system using one third of the recommended
minimum disk capacity by adding one or both of the first two
levels of support. The quality of the "pseudo paper tape" software
also makes it desirable to quickly boot in and out of the
operating system. By treating access to Network Central as access
to 64 separate devices we are able to overcome a severe operating
system restriction. Blocks in a file are stored contiguously in
OS/8 so that it is impractical for OS/8 to allow more than one
open output file on a device.

The second major operating system interfaced has been Digital
Equipments's RSX-11/M for the PDP-11. We have not attempted to use
Network Central to simulate the mass memory "systems device" but
have concentrated on contiguous file transmissions to gain access
to the shared peripherals and the CDC 6400.

The approach to be taken, and the desirability of mass memory
simulation is a function of the particular operating system to be
supported. After a certain level of sophistication is passed, it
may no longer be cost effective to simulate a mass memory device
used for swapping and general overlay support. As operating
systems become more complex, file manipulation, directory
processing and access permission code maintenance may force the

mount of local suport software to be excessive.

In all cases we have achieved substantial performance improvements at minimal hardware and software overhead. If more sophisticated systems than we have supported appear important enough for the additional effort, our approach would be to add an additional computer at the central site to provide the additional algorithms required for the several machines to be interfaced. We would adopt this more conservative approach over making substantial changes to the central machine because of the ease with which the interconnected network computers could provide the required support in a distributed fashion.

Distributed Computing

There are three applications areas which utilize the distributed capabilities of the network. First, a running CDC 6400 program can actively communicate with a program running either in a network computer or in a partition in our time-shared PDP-8. The network-based program generally acts as an intelligent terminal for the 6400 program. Experimental results are frequently transmitted to the 6400 after which the analysis is directed from the remote computer. Second, a new peripheral is often installed on a remote computer rather than the Network Central complex until its driver philosophy or final hardware interface is defined. Communication with a Network Central file or another remote computer is established to read from, or write to, the new peripheral. Finally, new program functions will often be programmed to run in a second remote computer which communicates with the first. In many cases the code to support a new function will either not fit or will degrade the original program performance below acceptable levels. In a recent analysis of program behavior, a small data taking routine was imbedded in a small time-sharing system. The data was logged at Network Central where a second computer read it, performing preliminary data reduction. In this instance, minimal change to the monitored program took place. The performance of the program was marginally affected by the data taking and far more sophisticated logging was possible because the initial data reduction routine was not forced into the already tight program being analyzed.

FUTURE DEVELOPMENTS
------ ------------

The system has reached a point of development during which the experience with current computers is being applied to interface new models. As figure 1 indicates, we expect to interface Data General Nova's, DEC PDP-11's, and a Raytheon 704. We anticipate each unique computer or operating system will require about

500,000 words of disk storage. At the other extreme we are adding some micro-computers.

One of the network computers supports a locally developed time-sharing system used for laboratories in computer organization and software fundamentals. We anticipate new application programs will be developed to support terminal, rather than computer, access to network resources.

The communication between Network Central and the CDC 6400 uses a transient program running in a computer in the central complex. The dialog with the 6400 is more complicated than that used in the remainder of the network. Parallel developments in the computing center and in our laboratory will result in effecting simultaneous communications between several network-based programs and corresponding programs running on the CDC 6400. The resultant communications discipline will be such that we expect to support telephone-connected computers, as well as the 6400, if we add a dedicated computer for these tasks.

CONCLUSIONS

The network has two key features: First, the universal serial interface with receiver control of character transmissions which minimizes interconnection hardware and software, and secondly, the fact that "intelligence" is distributed uniformly throughout the loosely connected remote machines. The result is an economical system which provides a marked increase in flexibility and accessibility without sacrificing essential real-time capabilities. The next logical step is to increase the connectiveness of idle machines to address the partitioning and distribution of large data processing tasks.

Our network design can be contrasted with networks of larger, self sufficient batch processing or time-sharing computer systems where load leveling or access to unique resources has been the aim, as a major function of our mini-computers continues to be real-time operation. Present intercomputer speeds allow us to relax somewhat the common sense restriction that data storage and program execution facilities reside in the same location. However, for real-time and on-line experiment controls, use of local storage is encouraged for guaranteed response time and data integrity. Using Network Central to store the only copy of a data file essential for achieving Nobel Laureate, tenure, or Ph. D. status is proscribed.

Within our mixed environment of real-time computing and computer system research, it is difficult to envision an approach which differs other than in detail from our present system. The two key

features of our network can be easily achieved by other
laboratories for much more powerful computing systems. Because our
experience emphasizes the desirability of initial homogeneity for
maintaining adequate control of the evolution to a heterogeneous
network, the existing population of mini-computers and their
operating system should be considered when designing an initial
configuration.

LITERATURE CITED
----------- -----

1. Barrett R.C. and Lennon W.J. "A Universal Serial Interface for
Data Set, Teletype, and Intercomputer Connection." May 1971 DECUS
Symposium Proceedings.

2. Lennon W.J., Barrett R.C., and Spies J.T. "Northwestern's
Mini-Computer Research Network." May 1972 DECUS Symposium
Proceedings.

3. Lennon W.J., Henderson K.F., and Spies J.T. "A Read Only Memory
peripheral for S-NET." November 1973 Decus Symposium Proceedings.

Note: DECUS Symposium Proceedings reprints are available from
either the author of the DECUS Headquarters in Maynard,
Massachusetts. In addition, the Proceedings of the Society are
available from University Microfilms.

APPENDIX

EXTRACT OF NETWORK USER'S GUIDE (Serial files)

All commands may be specified by the first two letters of the
command. The remaining letters until the first space character are
ignored. Exceptional, one letter commands are noted. All command
sequences must end with a carriage return.

NOTATION:

Character sequences a user must type are enclosed in single
quotes. Sequences generated by Network Central are enclosed in
double quotes. Optional sequences are bracketed. Parameters or
strings will be indicated by lower case words. When required,
non-printing characters will be denoted by three digit octal
numbers in parentheses.

IDENTIFICATION COMMANDS:

'LOGOFF'

This command will close out all account data taking for this terminal and (usually) cause the computer to halt. When a new user restarts the computer, instead of the standard "#" prompt character from the Network Central monitor, the local teletype will copy the following from Network Central:

"LOG ON!
#"

The user must then type in his password. Once logged on, the accounting procedures then maintain a log of all data transfers from that terminal and associate that users internal account number with all files created while using the computer. A users password is "secret" while the account number is public information.

SERIAL FILE MANAGEMENT AND GENERAL COMMANDS --

FILE NAMES:

A file name consists of up to ten characters followed by an optional '.' followed by two characters called an extension.

EXAMPLES: FROG   TEMP.PA

The extension helps to identify what type of file is there. With one exception, the extension is used for user information only. The ".IM" extension is appended to the program name a user types when issuing a "RUN" or "GET" command. Some commonly used extensions follow:

```
.SY     NETWORK SYSTEM FILE
.IM     NETWORK CORE IMAGE
.SV     OS/8 CORE IMAGE
.89     8090 BIOCTAL PROGRAMS
.PA     PAL SOURCE
.BN     NON-RELOCATABLE (LIKE PAL) BINARY
.FT     FORTRAN SOURCE
.DA     DATA
.LS     LISTING, PROBABLY TEMPORARY
.CR     CROSS REFERENCED LISTING, TEMPORARY
.TM     TEMPORARY
.FC     FOCAL
.BA     BASIC
.WU     PROGRAM WRITEUP
.TX     TEXT
.YF     WJL'S, PURGE THIS -- YOU'RE FIRED!
.WL     NOT AS SEVERE AS .YF ABOVE
.BS     PERSONAL, PROBABLY PHILOSOPHY
.01-.77 FILES LOCAL TO A TERMINAL
```

'ASSIGN' [file2]
'ASSIGN' [file1] '<'
'ASSIGN' [file1] '<' [file2]

Assign Network Central files, file1 and file2 as output and input files, respectively (the bracketed names are optional). Characters punched to a null output file are ignored while the end of information sequence, two ASCII CAN (230 octal) codes will be read from a null input file. If the "^" character appears immediately after the file name, that file will pack or be unpacked from the 12-bit word six bits at a time. Otherwise 8-bit packing is assumed, three 8-bit characters to every two 12-bit computer words. Since Network Central monitors all received data transmissions for the end of information sequence, files which might accidently contain that sequence are transmitted in 6-bit mode.

'DIVERT' [odev] '<' [file]

Divert the file to the queue of an output device. [odev] stands for a sequential output device and can be replaced by "LPT" or "PTP" under the current hardware configuration. An operator at the central site must initiate dumping of this file from the queue. The file must be explicitly "PURGED" following the actual output operation.

'RUN' (OR 'R') [program name]

Load the binary image file into core, release Network Central buffers and branch to the program's starting address. The assumed extension on the file is 'IM'. An explicit extension may be used to override the assumed extension -- E.G. ".89" for 8090 programs.

'GET' [program name]

Load the binary image file into core but do not branch to the starting address. Return to the network command mode to enter more commands before starting. When a series of "GET's" are typed in, the starting address saved by the network resident in the remote site is the last one received. The characters "/G" may be typed after either an assignment or "ASSIGN" or "OFF" command causing the Network Central Monitor to signal the remote resident to start the remote computer at the most recently received stating address.

'STATUS' argument

Where "argument" can be:
'FREE'
Print out how many free blocks (octal) are on the disk.
'CORE'

Print out how many 256 word buffers are left.
´USER´
Print the user (terminal) numbers for all computers allocated core buffers.
´DATE´
Print today's date.
´TIME´
Print the current time.
´TNUM´
Print out your terminal number (The one logged by the USER status command above).
´LOG´
Print out how many un-dumped blocks of the system log are on the disk.
´VER´
Print out the current version number.
´ACCT´
Print out the account number currently active at this terminal. Account numbers are public whereas the passwords are secret.
´FNT´ [character]
Included for completeness, this command is used only when performing "random access" input output with Net Central files.

´TALK´ [string]

This allows a terminal to send messages to the network central console TTY. An alert will sound at the central site until acknowledged there. [string] will then print out at the console. The remote site and console will then be "short circuited" to allow further communications. A control x terminates communication.

´OK´

Type this when a bulletin has been received to turn it off. Bulletins are entered at the central site and broadcast to all newly logged on users.

´NAME´ [file1]
´NAME´ [file1] ´<´ [file2]

The first form of this command creates an empty directory entry with the name file1. The second form renames the file, file2, to file1.

´OFF´ or ´O´

Release all Network Central buffers. This command should be used

to free up the two pages of core used for storing the command
string. The network program will either return to the system
monitor of the machine at the remote site or will halt.

    'PURGE' [file]

Remove all traces of the file from the network disk. If the file
is in use by any remote site or be on any output queue a "FILE
BUSY" diagnostic will be issued.

    'LINKS'

Make sure that the link block pointers to all files are now stored
on the disk.

    'MERGE' [file1] ',' [file2]

Concatenate file1 and file2 with null (000 octal) fill between.
File2 will disappear from the directory and the merged file will
be named file1.

MESSAGE TRANSMISSIONS:

Error messages and the line resulting from a 'STATUS' request are
preceeded by the message command character (004). They are only
one line long and are terminated by a carriage return and line
feed (215,212). They are then followed by the prompting sequence
which ends in the command character (005). For any given command,
the first character of possible error messages is unique. The
following is a list of all error messages:

    BAD DATE [Net Central only]
    BUSY FILE
    DEVICE BUSY [Net Central only]
    DISK FULL
    FILE NOT FOUND
    INVALID COMMAND
    INVALID DEVICE
    LOAD ERR
    LOG 75% FULL [Net Central only]
    NO!
    NO CORE (OOPS)
    NOT ACTIVE [Net Central only]
    NOT IN QUEUE [Net Central only]
    OUTPUT FILE EXISTS
    QUEUE EMPTY [Net Central only]
    SYNTAX ERROR
    SAME TO YOU!!
    TOO FEW ARGS
    TYPE 'DATE MM.DD.YY' [Net Central only]

6

A Computer Utility for Interactive Instrument Control*

PAUL DAY

Argonne National Laboratory, Argonne, Ill. 60439

I. INTRODUCTION

With few exceptions, laboratory automation has proceeded by the interfacing of a mini-computer to each instrument that requires some type of real-time service; data acquisition, experiment control and/or the analysis of an on-going experiment. To those willing to invest considerable programming effort for each system or those whose needs may be satisfied by a commercially available package, this approach appears attractive. However, limited system flexibility is the price paid for such systems, which price is derived from the inherent limitations of mini-systems; 16 bit word size, fixed system resources, custom tailored programs (not modular), inherent difficulty of program development on a teletype, and unavailability of the computer for experiment control during code generation. The recent trend toward interconnecting several minis together into a network overcomes the limitations of fixed system size and provides the possibility of generating new code while real-time service is being provided. However, there is a class of instruments which require, and others which could greatly benefit by, real-time support which included the storage and analysis of large amounts of data (25K bytes/sec.) using such routines as fast fourier transformations, multi-point smoothing, least squares analysis and

---
*Work performed under the auspices of the U. S. Energy Research and Development Administration.

matrix manipulation. This type of support cannot generally be provided by a mini-computer network unless a computer with significantly more capability than a mini is connected to the network. This computer should provide sufficient storage capacity and computational ability (32 and 64 bit floating point arithmetic) to permit the real-time execution of 20 to 30K word FORTRAN programs, and it could further increase scientific productivity if all the final analysis for each experiment could be performed on the same system.

Rather than a network, our approach is more analogous to an octopus, in which all real-time support is provided by direct connection to a central computer. A careful study of the real-time requirements of our Chemistry Division in 1967 indicated that a central computer with suitable hardware features, properly controlled by an operating system, could provide all the advantages of a large computer without sacrificing the isolation advantage of the mini-computer in the laboratory (1). The operating system must utilize these hardware features in a manner which will provide each user with the level of real-time and other service required to run his experiment, while at the same time insuring a level of data and program integrity enjoyed by the isolated mini-computer user. To date it has not been necessary to enhance our central computer's service by the addition of a mini-computer between the central computer and the remote instrument. However, some of our remote interfaces do perform computer-like functions which interact with an instrument on a microsecond time scale.

At first glance, the isolated mini-system has three advantages over a larger system (including a mini-network): isolation insures no competition for resources; data and program integrity are as good as the stand-alone software package; since hardware failure rate is somewhat proportional to the number of components in a system, the mini will be out of service less often. However, our system has demonstrated that it is possible to meet the above service and integrity aspects. The only disadvantage of our central system is a hardware failure rate of about once in 20 days compared to minis which often run for many months without failure. However, our on-line users consider this insignificant compared to the following advantages of our central system:

1. The system is of sufficient size and speed to perform most data analysis in real-time or at

the completion of an experiment. In the rare case that more computing capacity is required, 9-track magnetic tape is used for data transfer to a larger computer.
2. Dynamic sharing provides more resources (CPU, core, disk, magnetic tape) per experiment for a given capital investment.
3. Fast program generation and debugging with high speed card reader and line printer, along with the use of modular programming techniques makes program upgrading easy enough to keep up with the research scientist's changing needs. At least seven of our on-line experiments are considered state-of-the-art systems.
4. Time-shared execution of 25K word FORTRAN programs in real-time. Batch execution of up to 40K word programs for final analysis.
5. On-line keyboard displays and interactive graphics running at 9600 baud for working with data.
6. Disk storage (75 megabytes) for data (FORTRAN accessible), programs and program overlay storage.
7. Open-shop batch-processing permits on-line users to develop and debug their own FORTRAN programs which then have direct access to their data.
8. Only a small programming staff is required. A programming investment of 15 man-years has completed the operating system and applications programs for 21 on-line experiments. An additional 8 man-years was invested in the design and construction of the hardware interfaces.

All control and organizational aspects of the System were designed, coded and implemented by our Chemistry Division (2). The following Xerox software was interfaced to the system; loader (BCM), assembler (SYMBOL), compiler (FORTRAN IV-H) and FORTRAN run-time/arithmetic subroutine package. The central facility has been self-running (no computer operator) for over five years 24 hours/day, 7 days/week with the exception of 8 hours: 4 hours of preventive maintenance followed by 4 hours of System development each week. The system is continually evolving to support new peripherals (disks, interactive terminals, array processors, etc.) and to provide new services as the needs arise. Provisions have been built in for automatic recovery from power failures, to restart faulty peripherals and to cease using faulty memory when detected. System uptime averages 157 hours per week with one system crash (down time averages about 40 minutes) about every 10 days (half are hardware related and the other system software). Automatic restart for users is available after weekly maintenance or a crash. Daily copying to magnetic tape of all program and data files stored on the disks and RAD (Rapid Access Device - fixed head per track disk) insures users against catastrophic failure. A partial loss of data collected over several hours has occurred three times in five years as a result of hardware failure.

II. System Services

Computer Configuration. The present computer configuration (Table I) has been incrementally expanded from a 24K word system, without disk, supporting 8 experiments to the present system supporting 21 experiments and six independent graphics/keyboard terminals.

Sigma 5 CPU	Central Processing Unit/ floating point arithmetic
Memory	56K 32 bit words plus parity
MIOP (2)	750K bytes/sec. bandwidth, 32 channels
CIOP	21 110 baud teletype, 2 1200 & 8 9600 baud terminals
Disk	251K bytes/sec. transfer, 25 megabytes x 3 spindles
RAD (fixed-head disk)	170 Kbyte/sec., 6 megabytes
9-track mag tape (2)	60 Kbytes/sec., 800 bpi
Card Reader	1500 cpm
Line Printer	650 lpm
Card Punch	300 cpm
Paper Tape	reader: 300 cps Punch: 120 cps
Digital Plotter	CALCOMP 565

TABLE I. Hardware Configuration

Data Transfer. All user and system data is handled by input/output hardware (MIOP) that interacts with all devices running concurrently; being capable of transferring data at an aggregate rate of 750K bytes/ sec. Thus, a device seldom waits more than a few microseconds for transfer of a datum after it signals the MIOP that it is ready. Each user's data is moved directly between his assigned program core memory and a device; there is no system overhead required beyond the time required for MIOP transfer.

Instrument Control

Real-Time Response. A software priority is associated with each program controlling an interfaced instrument. These priorities are assigned on the basis of instrument requirements at system boot-in time. Upon receipt of a request for program execution (e.g. user's device signals "done"), the user's real-time program will commence execution after a time lapse that depends on its priority. The highest priority program will commence execution within 160 microseconds, whereas the lowest priority may not start execution until up to 200 milliseconds later, if one or more higher priority programs are currently "ready-to-run". A simplified real-time execution service cycle may be represented by the following:
1. Program requests a system service (e.g. start reading data from my device)
2. Program calls end-of-service (EOS) (program has nothing to do until data transfer is complete)
3. An MIOP interrupt for this device indicates transfer complete
4. Program processes data
5. Proceed to step "1"

Custom Designed Control Programs. Even though our engineers have established a basic interface design and our programmers have developed effective techniques for acquiring data and controlling instruments, each new system is custom designed at a level which will best serve the user's current and projected needs. A command language is established with which the user can easily communicate with his programs in a natural manner. Program prompts are provided (e.g. "how many scans") so the user does not have to memorize an unfamiliar mneumonic scheme. While our programming staff designs and writes the data acquisition and instrument control portions of the package, users familiar with FORTRAN may code and use their own FORTRAN programs for real-time processing and/or final analysis.

User Communication and Control. Human communication with the computer system is provided at each remote experimental site via a KSR33 teletype. At system boot-in time, input requests are initiated to all communications terminals. At any subsequent time, a user may log-on to the system; at which time sufficient core memory space is assigned for the core resident portion of his specific instrument control program and interact with his instrument through this program. Some of the remote sites also have interactive graphics terminals. In addition, several graphics terminals and keyboard displays are strategically located for shared use. A user may similarly log-on and interact with his graphics and/or computation programs from these terminals.

The system provides the user with complete control of his instrument interface through the construction and execution of any I/O command sequence that can be executed by the MIOP and his device. This permits the buffering of data into a user's core area, with interrupt indications being passed to the user's control program as each buffer fills, while the MIOP may initiate further data transfer into another user-specified area within less than 10 microseconds after a buffer fills. System integrity is insured by checking for core bounds violation before an I/O operation is initiated on the user's assigned device.

Mass Storage Usage. The system was designed to maximize data throughput. Therefore, disk (and RAD) file structures were simply designed to provide random access in record sizes of the user's choice. When defining a new file, the user provides the name and size of the file to be defined. The system assigns a block of contiguous sectors and records the file name, location, size and creation date in the user's disk-resident file index. The file may be "opened" at any subsequent time, which then places the file's sector bounds in the requesting level's PDT (program description table). The file space within these bounds may then be used in any manner the user chooses; however, data transfer always commences at the beginning of the logical disk sector (LDS) specified (one sector = 256 words). Each request contains the LDS, the core address for data transfer and the number of words to transfer. The actual starting sector address is then derived by the addition of the LDS to the lower sector bound (core-resident after file is "opened"). After insuring that the user's core and file bounds will not be exceeded, the operation is initiated.

To facilitate the rapid updating of large histograms, only a portion of which can reside in core memory at one time, a write-read feature has been provided. This permits the writing back on the disk (or RAD) of a just-updated portion of a histogram and the reading into the same core area of the adjacent portion of the disk-stored histogram; all in one disk rotation.

Non-Resident Program Execution. Our real-time computational requirements vary over a wide range. Our pulsed NMR experiment requires scan averaging a 16K word histogram every 300 milliseconds, taking about 100 milliseconds per update. While this is the highest usage required by this experiment, it is needed occasionally. Other experiments require the execution of a 25K word histogram transformation program (nuclear fission detectors) every minute, taking about 5 seconds. Other users require this type of execution every 10 minutes for times ranging from a few seconds to 30 seconds.

To satisfy this variety of demand without requiring an inordinate amount of core memory, the operating system provides for the time-shared execution of non-resident programs (not always in core) in the background core area. These programs are disk-resident core images of relatively large programs required infrequently and without severe time constraints. Two queues are provided: one with a 1 and the other with a 32 second execution time limit. These programs are usually FORTRAN which are user-written or from user-group libraries. This system function also makes it easy for the casual user, who is conversant with FORTRAN, to perform his own real-time processing.

Prior to execution, a non-resident program must be "opened", which consists of creating and storing the core image on a disk file and setting the relevant file information in the user's PDT. Multiple programs may be opened and/or open at one time. At execution request time, using the core resident size indication, the system writes a sufficient portion of the current lower priority task on the disk and then reads in the core image non-resident program. The total time required between the request and the start of non-resident execution is about 250 msec. for a 15K word program, if no other request is pending; otherwise it must wait its turn in the queue. Several options are available at execution time; core page size if larger than the code, automatic "opening" of user's current foreground files to the non-resident program, queue selection (1 or 32 sec. time limit) and a "save" option (write program back on disk at exit time). The "save" option allows step-wise execution. At entry time, a register contains the address of the requestor's calling function parameter table (which may contain an argument list of any length) facilitating argument transfer. Non-resident memory write locks are "ORed" with the requesting level's write locks, permitting modification of the foreground core.

<u>Concurrent User Operations</u>. A user's program has the option of overlapping different system services. This is effected through a "type" option in each call a program makes to the system, specifying the type of

action the system should take after initiating the request:
1. Initiate request and return immediately
2. Initiate request and return when request is completed
3. Return if or when the previous request of this kind is completed
4. Initiate request, call end-of-service and return when request finished
5. Initiate request, return immediately and return when request finished

Multiple "request finished" indicators are saved in the requestor's PDT, with user notification being provided on a first-in-first-out basis. Using combinations of the above "types", a user program may overlap a number of operations that do not have to be completed in a specific order. For example:
1. Start device reading into next data buffer
2. Move previous data to magnetic tape
3. Write data to disk
4. Execute a non-resident program which generates display data
5. Computation

Having the option of being notified about the completion of the various processes, the user program can proceed as required.

Graphics Facilities. Graphics facilities consist of a CALCOMP 565 digital plotter and eight Tektronix 4010 graphics units, three of which have 4610 hard copy units. Their respective FORTRAN callable driver subroutines write calculated move-data to the user's disk plotter-file. After executing the FORTRAN program in the batch mode, a real-time CALCOMP utility program is initiated, which reads the move-data from the disk file and writes this data to the MIOP-coupled CALCOMP. Thus the slow moving CALCOMP plotting is performed while other batch jobs are running.

The Tektronix interactive graphics units are supported with driver routines that execute at a foreground level which in turn makes calls to the user's non-resident code, return to the foreground level reads the move data from the disk and transmits it to the Tektronix screen. Or, the foreground writes an inquiry on the screen and, after a subsequent user response, returns the response to the non-resident program via a new execute non-resident program request. Thus, non-resident core is available for other users during writing, reading or waiting for a user to respond to an inquiry. Provision is also made for the user's program to ascertain the coordinates of thumbwheel controlled cursor crosshairs, providing the user further interaction with his program.

Time-Sharing. Previously compiled and stored (batch mode) FORTRAN IV-H programs can be executed in a time-shared mode from any remote keyboard terminal (numbering 31). The keyboard terminal is treated as the I/O device in place of the card reader and line printer. Actual code execution takes place in background core as a non-resident program between I/O statements. Programs executed in such a manner also have access to disk and RAD storage. In addition to the users' specialized programs a library of conversational programs is available. Routines for interactive graphics terminals include: peak resolving, data smoothing, spectral displaying. Interactive routines are also available for manuscript generation using a text-editing package.

Open-Shop Batch-Processing. A very important feature of the system is the support of an open-shop batch-processing service in which jobs are queued through the card reader. This type of queuing provides the casual user with immediate feedback for the rapid debugging of his programs. Batch is the only level in the system which has access to the "batch peripherals"; card reader, line printer, card punch and paper tape reader/punch. The batch level also has access to disk, RAD and magnetic tape. Although the on-line user has the option of performing extensive analysis of his experiment from his remote terminal, the batch level is often utilized where large amounts of output are involved and for the transferring of file data onto magnetic tape for archiving or transport to another computer. The batch level is also used extensively to generate and debug code for the control of on-line experiments.

The batch level uses all CPU cycles not used by on-going foreground processes while no higher priority use (e.g. non-resident execution, foreground loading, core image generation) is required of the background core area. Under normal daytime system loading, the foreground users require about 10 percent of the CPU cycles and the non-resident program execution about another 40 percent. Thus, it appears to the batch user that his program is executing on a computer with about half the speed of a Sigma 5 computer which was dedicated to batch-processing.

Long-Term Computation. CPU utilization seldom exceeds 30 percent in a 24 hour period even with heavy real-time and batch usage. The remaining CPU cycles

are made available for executing very long (hours to
weeks) batch-type computations running at a priority
level below batch processing. These jobs differ from
normal batch in that they only have access to disk and
RAD files, not the "batch peripherals". Once initiated
(from the card reader), the job is rolled into background core anytime there is sufficient core space and
higher priority usage permits. The daily saving of
files onto magnetic tape also copies the current core
image (now stored on disk) of the long term job along
with its disk and RAD files. Automatic file (and long-
term) restoration at system boot-in supports executions
extending over long periods.

<u>Incremental Expansion</u>. As more experiments are
added to the system, one is concerned with two types
of system expansion; hardware and software. The Xerox
hardware architecture is such that essentially all
hardware capability is modularly expandable except the
CPU. For example memory can be expanded in increments
of 8K words to a maximum of 128K; I/O channels can be
added in groups of 8 up to 64 (2 MIOPS) until another
memory port is added which would then permit up to
another 64 channels. The addition of a new interfaced
instrument requires the physical connection of its
associated device controller to the MIOP bus structure,
a one hour job, during which computer power is shut
down.

Software expansion need only be concerned with the
generation of applications programs for each newly interfaced instrument. The operating system does not
require any modification for a new experiment. At
system boot-in time, all of the required PDT's are
generated from level parameter cards, one per interface.
Therefore, a new instrument requires the addition of
only one card to the boot-deck. The applications
programs for a new instrument are designed and written
in close consultation with the users while the new
interface is being designed and constructed. Total time
from conception to operation runs about 3 to 6 months.

III. <u>Current On-Line Experiments</u>

The 21 experiments presently interfaced to the
Sigma 5 require widely differing types and amounts of
service. Data transmission rates vary from one byte
per minute to 100,000 bytes per second over short
intervals. Control requirements vary from none, to
computing the angular coordinates of a goniometer from
a crystallographic orientation matrix, to providing
pulse sequencing for radio frequency generators. Mass
storage requirements vary from 1000 to 1,000,000 words.

6. DAY   *Instrument Control*   95

Real-time processing includes; scan averaging 16K
channel spectra, performing mass-energy transformations on correlated pulse-height events, fast fourier
transforms, least squares analysis, data smoothing and
peak finding. Table II is a list of the currently
supported programs with their associated instruments.

Quantity	Research Program	Instrument Type
5	fission and nuclear spectra	Packard 45 pulse height analyzer
2	routine chemical analysis	mass spectrometers (12 in. 60° sec)
1	pulsed radiolysis	Biomation 8100 transient recorder
1	stopped-flow kinetics	Biomation 802 transient recorder
1	precision heat capacity	Vidar 5204 data acquisition system
1	crossed molecular beams	time-of-flight detector
1	crystal structures	neutron diffractometer
1	biological system structures	Varian HR220 super-con pulsed NMR
1	very fast kinetics	Nuclear Chicago Multiscalar (400)
1	chlorophyll studies	Cary 14 spectrophotometer
1	matrix isolation	Cary 17H spectrophotometer
1	30 ft. grating spectrometer	plate measuring comparator
1	primary photosynthesis	Varian E-9 ESR spectrometer
1	plant pigment studies	Varian E-700 ENDOR spectrometer
1	environmental radioactivity	multi-detector (22) counting system
1	high-temperature chemistry	effusate spectrometer

TABLE II. Current Interfaced Instruments

Pulsed Nuclear Magnetic Resonance Spectroscopy.
This system consists of a Varian HR220 super-con NMR
operating in pulsed mode for producing fourier transform ^1H and ^{13}Cmr spectra (3,4). Automation is provided to control pulse sequencing with up to 10
interval-width pairs: pulse-widths of 0.5 to 511
microsecond in 0.5 microsecond steps and intervals of

M x 2^N (where M and N range over 0 to 31). Data is digitized with an ADC (8 to 13 bit selection) and transmitted directly to computer memory in groups of up to 16,384 channels with dwell times of as low as 16 μsec without losing more than 5 percent of the scans. Data loss is hardware detected and software compensated; incomplete scans are discarded. For shorter dwell time a hardware/software interlace feature has been provided which takes every N-th (N = 2, 4, 8, 16) data point from N successive scans in an advancing manner, requiring up to 16 times longer accumulation periods, but permitting acquisition of spectra with dwell times as short as 1 μsec. Data is scan-averaged in real-time onto a disk-stored histogram. After accumulating the desired number of scans the data is base-line corrected, apodized, fourier-transformed (4K channels requires 3 sec.) and phase corrected. Spectra may be stored for future comparison or analysis. A "tau" mode is provided which will automatically collect and analyze spectra at up to 10 specified intervals. The spectrometer is primarily used for natural products chemical studies, especially in photosynthetic pigments and in electron transport proteins.

Neutron Diffractometer. We have an extensive program investigating crystal and molecular structures in the 10-300°K temperature range on the following types of materials: hydrogen bonded such as hydrated protons; high conductivity inorganic complexes; bonding electron density studies on inorganic and biological compounds. Sigma 5 support for this effort includes automation of a four-circle neutron diffractometer (5,6), located one mile from the computer, which is controlled at any level of interaction selected by the experimentalist. This interaction can be either at the reactor or in the user's remotely located office (not at the reactor). User commands range over the following: move circle "X" to N-degrees, take a count, scan a diffraction peak in a specified region and store its coordinates, compute a preliminary orientation matrix, determine the locations of a set of reflections using the preliminary matrix, compute a least-squares adjusted orientation matrix using selected reflections, scan reflections (up to 10,000) in selected regions of reciprocal space. These operations may extend over periods from minutes to weeks (the program is automatically restarted at the appropriate point after system shut-down or crash) without human intervention or they may be suspended, discontinued or terminated at any time. With typical crystal sizes,

about 150 to 400 reflections are scanned per day. Provision is also made for real-time interactive editing and graphic display of data.
Analysis of structures containing up to 50 atoms in the unit cell are generally run in batch mode. The analysis includes usage of: the Canterbury fourier program for obtaining 3-dimensional fourier analysis of intensity; ORFLS-3 least squares routine with isotropic temperature factors; ORFFE-3 for obtaining atom distances, angles and dihedral angles. These programs may be run during the data acquisition phase to assist in establishing the measurement strategy of a particularly difficult crystal system.

Pulse Radiolysis. A broad range of pulse radiolysis experiments is conducted at the "LINAC" (electron accelerator) located about 1200 feet from the computer. They have developed a new method to record the time dependence absorption or emission spectrum produced by a single pulse of electrons. The time resolved spectrum is produced by an image converter camera with a streak capability (7). The streak image on the phosphor, which is a two dimensional array of the absorption (or emission), is scanned by a TV camera and stored on a video disk. Under computer control: the video disk image is scanned one line at a time; the scan is digitized and stored as 2000 points in a Biomation transient recorder; the Biomation data is then transmitted to a disk file on the computer. This process is repeated until the data from about 100 scan lines has been stored. The data from one scan line represents the light intensity at one wavelength as a function of time. The experimenter may view this data and interact with his analysis programs via a Tektronix graphic display unit, which has a 4610 hard copy unit. After further processing, a FORTRAN program is used to generate a 3-dimensional display of intensity versus wavelength and time. Alternate programs are available to give 2-dimensional plots of intensity versus wavelength. The entire process of obtaining the data and storing in the Sigma 5 takes a matter of minutes compared to days by the conventional methods and makes working with highly radioactive materials more feasible.

In measuring fast kinetics (50 picosecond risetime) which are initiated by a pulse of electrons, signals are often small and signal-to-noise ratios are limited by shot noise, which means a great deal of signal averaging may be necessary. The computer system allows the saving of partial averages, so that long averaging times may be used without fear of losing everything caused by an experimental problem in the last five minutes. Good partial averages are averaged and kinetic data are extracted using non-linear least-squares curve fitting; editing and display are provided on an interactive graphics terminal located at the experimental site. This scan averaging system is effected through an interfaced 400 channel Nuclear Chicago multi-scaling analyzer.

ENDOR and ESR Spectroscopy. A significant portion of our efforts to characterize primary photosynthesis is being supported by two interfaced spin resonance spectrometers (8,9): a Varian E-700 ENDOR (electron nuclear double resonance) and a Varian E-9 ESR (electron spin resonance) buffered by a Nicolet 1074 CAT (signal averager). The uniqueness of these spectrometer systems derives from the large memory and computational speed (floating point arithmetic) of the Sigma 5 which allows the acquisition and manipulation of large quantities of three dimensional data (3-D) via two ADC's. A single 3-D experiment may be composed of up to 640,000 data points. For example in the ENDOR mode signal intensity and radio frequency are recorded as a function of magnetic field. In the ESR mode, the typical intensity versus field spectrum is recorded as a function of a third dimension, such as time for kinetic studies, or light intensity for photochemical studies. Furthermore, spectrometer operation is controlled by the computer; for example, in the ENDOR mode the computer synthesizes the radio frequency value, makes decisions about data validity, and terminates spectrometer operation if the radio frequency power is insufficient. All data calculations, including sophisticated spectrum simulations as well as FFT (fast fourier transform) analysis is performed in real-time by user-written FORTRAN programs. These unique facilities have greatly simplified and speeded up the unraveling of various fundamental aspects of primary photosynthesis in our laboratory.

Stopped-Flow Kinetics. Using the stopped-flow technique, kinetic studies are being made on systems with short-lived reactants and products at a quantitative level never before achieved (10,11). Current studies cover a variety of actinide redox reactions, including work with Np(VII), one of the most powerful oxidants known in acid solution.

The apparatus consists of a Durrum model D-110 stopped-flow spectrophotometer whose output is digitized and stored in a Biomation 802 transient recorder, which in turn is interfaced to the computer. One thousand data points spanning a time range of 0.5 msec. to

20 sec. are collected per experiment. Real-time storage
and analysis of the data allow output of the least
squares fit results at the remote site within 30 sec.
of the end of the experiment. The fast turn around
allows the experimenter to adjust conditions continually for optimum results.

Spectroscopic Plate Reading. The spectroscopic
group is using a 30 foot grating spectrometer to inter-
pret actinide spectra, each of which may have up to
100,000 spectral lines. An automatic comparator
measures the position of spectrum lines on an exposed
photographic plate by moving it continuously under a
scanning slit and photocell; the local blackening at
selected intervals is obtained by digital voltmeter
(DVM) readings of the photocell current. About 60,000
readings are taken on one scan of a 250 mm. plate,
requiring about 4 minutes. The digitized (DVM) data
is transmitted directly to the computer, stored and
analyzed in real-time using non-resident execution.
The smoothed readings and derivatives are scanned for
peaks, and the peak positions are converted to wavelength by a polynomial formula derived in a separate
scan of known standard wavelengths. An important convenience is a disk file of standard wavelengths and
positions; only one pair of knowns is required as input
calibration, so the system requires much less effort
than manual measurement and at the same time gives
increased accuracy (1 in 10^7).

Nuclear Particle Detection. A broad area of research effort in our Chemistry Division is concerned
with eliciting the details of the fission process. In
studying the dynamics of a multi-nucleon fissioning
system, it is necessary to measure the changes in the
fragment mass and kinetic energy as the initial excitation energy, angular momentum, and target mass and
spin are altered. The initial conditions are most
easily varied by using direct-interaction induced fission reactions such as (d,pf) which allow the simultaneous observation of excitation energies from below
to 10 MeV above the fission barrier. Characterizing
these reactions requires a five parameter event-by-
event correlation of the fission fragment energies as
well as the light particle type and energy. These
quantities are converted to mass and kinetic energy
distributions in real-time by user-written programs
(executing as non-resident programs), permitting sufficient interaction between the experimentalist and a
real-time display of his transformed data; facilitating
optimum use of the experimentalist's and accelerator
time.

The computer system supports five data acquisition areas located in the Chemistry building, Cyclotron building and at the Van De Graaff accelerator in the Physics building located about 2,000 feet distant. Experiments range from single pulse height spectra (using Packard 45 memories and model 160 ADC's), recording of event-time data (at resolutions of 40 μsec.), to multi-particle pulse height spectra on up to six ADC's. Singles, correlated and transformed spectra are generated in real-time and automatically displayed at a selected interval. An option is also provided for storing event data pulse-heights (one or more ADC's) on magnetic tape for subsequent analysis in batch mode or on to the disk for real-time display, editing and analysis. Event rates of up to 10,000 ADC-events/sec. may be processed.

UV, Visible and Infrared Spectroscopy. Two interfaced Cary spectrophotometers (14 and 17H) support the Division's chlorophyll and surface studies groups. Digital data from the Cary 14 is sent directly to the computer for subsequent deconvolution and plotting on the Calcomp plotter. On the Cary 17H, the scan speed, chart speed and intensity range are under computer control (12,13). Direct digitization of the photomultiplier output has enhanced the absorbance accuracy by at least a factor of two. Automatic absorbance scale ranging is also provided during a spectral sweep such that the useful absorbance range has been increased from 0.3 to 0.4. In addition, the problem of data fidelity depending on scan speed has been eliminated. Repetitive scans with data averaging provides a ready means of improving signal-to-noise ratios. The digitized data are available for analysis by programs which include base line correction, Gaussian-Lorentzian deconvolution, peak area measurement, and plotting of the original or program-modified data.

Heat Capacity Measurements. This system is used for the precise measurement of the heat capacities of various chemical compounds between 0.1°K and 350°K, a platinum resistance thermometer being used between 4°K and 350°K and a germanium resistance thermometer between 0.1°K and 25°K (14,15). Thermometer resistances are measured with a 6-digit potentiometer whose dial positions are automatically transmitted to the computer on command. A fifteenth order polynomial (for germanium) which converts voltage to temperature is solved by a non-resident program, returning the time, temperature and temperature drift. After drifting becomes constant, a heating cycle is entered in which readings of heater power and time are read automatically by an interfaced Vidar 5204D-DAS and trasmitted to the computer. Following heating, the thermometer current and resistances are again followed, until drifts become constant. The system makes the recording of the data much easier and more reliable and in addition it is much easier for the experimenter to decide when equilibrium has been attained after the heating period, so that another heating can be initiated.

IV. New Directions

Our central computer system has supported present-day instrumentation quite well. However, as the art of automation advances, experimentalists' learn to better utilize their new sophisticated tools and new instrument capabilities evolve, it is clear that new and more difficult data acquisition problems will continue to arise. One of the more demanding problems is the scan-averaging of spectra acquired at rates approaching a million channels per second. Another is more sophisticated real-time control of experiments requiring logic decisions (computation) on a microsecond time scale. One might try to solve these problems by adding a mini-computer between these new instruments and the central system. However, one is then faced with the problem of programming these individual computers (probably of different manufacture) and designing an interface to the central system. In addition, a mini-computer's response time is faster than a central computer only by virtue of the fact that it takes 4 nanoseconds per foot (2 way transmission) for data transmission and communication.

We are intending to provide higher speed capabilities in the future by using microprocessor support at the experimental site. Xerox has a System Control Unit which has a wide variety of configurations and options including 8 registers, interrupts, stacks, clocks, read-only and writable memory (16 bit, 350 and 700 nanosecond) in increments of 1K words, and optional floating point. We will load these computers with code from the central computer for each phase of its data acquisition effort, using them more as a sophisticated input/output device than as a computer. This will keep the programming logically simple and thus efficient. In addition, this type of usage will not require much memory, thus keeping the cost down below that which would be required by using a mini-computer.

V. Summary

The successful operation of our system over the past five years is a clear demonstration that a central computer, without remote mini-computers, can provide very sophisticated real-time support without suffering the disadvantages of unreliability often experienced in large systems. Users never experience the consequences of another user's errant program, only their own. The level of system integrity is such that we test a completely new instrument control program during normal system loading, without having to be concerned about interferring with any other user's interests. Evidence of the system's acceptance as users have been added over the years, is that we continue to have 6-8 new users wanting to be interfaced as funds become available. The total capital investment for the entire system is about $50K per experiment, which compares favorably to the cost of sophisticated mini-systems. But, one has much more resource and capability to support each experiment: providing full experiment control in a flexibly interactive manner; fully interactive graphics capability with on-going and completed experiments; computational ability for complete final analysis of experiments.

APPENDIX

Memory Bus Structure. The 960 nanosecond core memory (32 bits plus parity) is modular; each 16K words has its own read-write and address recognition capability and may have up to six ports. The CPU and each processor (MIOP's and CIOP) is connected to its own "port bus" made by joining one port from each memory bank. Simultaneous access is achieved when requests are in different banks and cycle stealing takes place when more than one request coexists for a single memory bank.

Memory Protection. A 2-bit lock and key feature prevents a user's program from modifying unassigned core, while permitting the execution of reentrant code. The "key" resides in the current PSD (program status doubleword), the page (512 words) locks are set in fast memory in the CPU via a single instruction by the scheduler each time a program level commences execution. Implementation of the protection feature will trap a core-modify instruction before execution and yet does not increase instruction execution time.

6. DAY Instrument Control

Data Flow. Each MIOP (multiplexor input/output processor) may handle an aggregate data flow of 375K memory accesses per second (one or four bytes) on up to 32 concurrently active data channels (typically one per experiment). Only a single instruction must be executed by the CPU to initiate a data transfer between any core memory area and a specified device; all subsequent data movement and communication with the remote device is handled by the MIOP. At the termination of data transfer (up to 65,536 bytes), the MIOP generates a single interrupt. Upon recognition of the interrupt, the CPU ascertains the channel number and the channel-end conditions by executing a single instruction. Thus, high speed data transfer to individually selected core areas (specified by each user's program) is completely handled by the MIOP, freeing the CPU to perform useful computations concurrently with data transfers.

Terminal Communications. The CIOP (communications input/output processor) handles up to 128 lines of mixed speeds from 110 to 9600 baud; currently driving 21 teletypes, 8 interactive graphic units and 2 keyboard displays. This processor works with the CPU in a manner similar to that used with the MIOP. Thus message transfers take place to any core memory area while the CPU is performing computations.

Experiment Interfaces. Each of the remotely located on-line experiments is interfaced to a remote controller. The simplest controllers consist of a one byte buffer and a few logic lines. The more complex ones have up to sixteen byte buffers, computer initiated timing circuitry to control instrument functions to a precision of 0.1 microsecond and numerous logic lines. For transmission rates up to 100,000 bytes per second at distances up to 1500 feet, a remote controller is connected to a device controller via a bundle of co-axial cable (8 bit wide data path). For transmission rates of up to 4K bytes/second and at distances of up to one mile, four twisted pair are used. The device controllers were designed by Argonne engineers and are connected to the MIOP bus structure in the computer main frame, weekly data transmission validity tests on all interfaces running concurrently for about an hour never detect a transmission fault. About every eight months one of the interfaces fails solidly because of component failure, which is readily fixed within a few hours.

Core Memory Assignment. With the exception of system areas, core memory is assigned dynamically. As users log-on from their terminals, sufficient core is assigned for the resident portion of their data acquisition program, which includes their own data buffer areas. The programs are structured into user-controlled overlays which are typically written onto their disk file and rolled into their core area as they require. Using this overlay technique, the average core residency required per program runs between 2 and 5 pages (page = 512 words). To perform real-time computations using FORTRAN, the foreground code may call for execution of a "non-resident" program. Foreground core area is assigned from the top of core down in the first available block of continguous core pages. Background core (batch and system tasks) is assigned from lower core upward. This permits the maximum background core area to extend on up to 42,600 words when no real-time tasks are loaded. While background core is variable in size, the user is always guaranteed an area of at least 15,900 words. Real-time job lengths vary from minutes to weeks. On a typical day, about 60% of the instruments are on-line at any one time, requiring about 14,400 words of core. This leaves about 28,200 words of background core.

Reentrant Code. The Xerox write-lock feature, which permits the execution of code anywhere in memory yet forbids the modification of memory without the proper "key", facilitates the utilization of reentrant code to perform many user functions. Reentrant code is defined as code that does not modify itself. About 800 words of reentrant code are in permanent residence for performing such functions as two-way translation of keyboard I/O and binary-coded-decimal conversion for some of the interfaced multi-channel analyzers. In addition, there are 128 words of EBCDIC-ASCII conversion tables available.

Program Description Tables. In providing a multi-programmed environment for a large number (currently 41) of concurrently running programs (user and system), it was found that a program description table (PDT) associated with each priority level was an efficient method of keeping track of system service requests and individual program status. Besides simplifying priority queuing and level termination at end-of-job, a PDT structure also simplifies the addition of a new program level to the system. PDT generation is controlled at system boot-in by one level-parameter card per terminal.

The PDT for each priority level is 80 words long plus two words for each I/O command pair needed by the level (ranging from 1 to 16) to control its associated device. The table content also includes; user ID number, level status, program status double words (PSD), memory write-lock image, time remaining, core bounds, disk memory file bounds and register values. These tables are stored in a write-protected area of core memory.

Scheduling. The system structure/operation and usage strategy have been developed concurrently. Thus a system has evolved which maximizes the system work per unit time in such a manner as to provide the required level of real-time interaction between the user, his keyboard and his instrument. However, the specific strategy used for any particular application is not dictated by the system as long as programs make requests which conform in time, core memory, device operation and file bounds. A hardware trap is set to disallow the execution of privileged instructions by user code.

The basic scheduling algorithm runs each task to completion, contingent upon its priority. The scheduler determines what priority level will execute next, through the processing of all I/O interrupts from the MIOP, and the maintenance of software priority level status using the PDT's. Once the identity and status of the level associated with an interrupt is determined, the scheduler compares its priority with that of the level interrupted. If it is of lower priority, the interrupted level is resumed. Otherwise, the execution dependent portion (PSD, execution time and registers) of its current state vector is moved to its PDT. Referencing the new (interrupt signalled) level's PDT, its execution is initiated by setting the memory write-protect locks, execution time, registers and the PSD.

Upon receiving an end-of-service (EOS) request from a level, the scheduler records this in its PDT. Then it scans down the priority chain through the PDT's until it finds another level that is "ready-to-run". State vectors are then adjusted as above to initiate execution.

A necessary feature in a multi-programmed system is the prevention of processing lockout by some higher priority level looping endlessly. A maximum service cycle time (execution: foreground=1 sec., non-resident= 1 or 32 sec., batch and long term=no limit) and a time remaining value are stored in each PDT. Any time a level is executing, the time-remaining value is decremented by a CPU clock (2000 Hz). But each time a level

calls EOS, the maximum service value is moved into the
time-remaining PDT location. By breaking processing
into logical service cycles, a program can readily
perform its function forever, but will be terminated
by time-out if it malfunctions time-wise.

ABSTRACT

A careful study of our laboratory automation needs
in 1967 led us to the conclusion that a central computer could support all of the real-time needs of a
diverse collection of research instruments. A suitable
hardware configuration would require an operating system to provide effective protection, fast real-time
response and efficient data transfer. An SDS Sigma 5
satisfied all our hardware criteria, however it was
necessary to write our own operating system; services
include program generation, experiment control real-
time analysis, interactive graphics and final analysis.

Our system is providing real-time support for 21
concurrently running experiments, including an automated neutron diffractometer, a pulsed NMR spectrometer
and multi-particle detection systems. It guarantees
the protection of each user's interests and dynamically
assigns core memory, disk space and 9-track magnetic
tape usage. Multiplexor hardware capability allows the
transfer of data between a user's device and assigned
core area at rates of 100,000 bytes/sec. Real-time
histogram generation for a user can proceed at rates of
50,000 points/sec. The facility has been self-running
(no computer operator) for five years with a mean time
between failures of 10 days and an uptime of 157 hours/
week.

Literature Cited

1. Day, P. and Krejci, H., Proc. AFIPS FJCC (1968)
   33 1187-1196
2. Day, P. and Hines, J., Operating Systems Review
   (1973) 7 (4) 28-37
3. Scheer, H. and Katz, J. J., Proc. Natl. Acad. Sci.
   USA (1974) 71 (5) 1626-1629
4. Crespi, H. L., Kostka, A. G. and Smith, U.,
   Biochem. Biophys. Res. Comm. (1974) 61 (4) 1407-
   1414
5. Peterson, S. W., Willett, R. D. and Huston, J. L.,
   J. Chem. Phys. (1973) 59 (1) 453-459
6. Williams, J. M., Petersen, J. L., Gerdes, H. M.,
   and Peterson, S. W., Phys. Rev. Lett. (1974) 33
   (18) 1079-1081

7. Gordon, S., Schmidt, K. H. and Martin, J. E., Rev. Sci. Instrum. (1974) $\underline{45}$ (4) 552-558
8. Norris, J. R., Uphaus, R. A. and Katz, J. J., Chem. Phys. Lett. (1975) $\underline{31}$ 157-161
9. Norris, J. R., Druyan, M. E. and Katz, J. J., J. Am. Chem. Soc. (1973) $\underline{95}$ 1680-1682
10. Watkins, K. O., Sullivan, J. C. and Deutsch, E., Inorg. Chem. (1974) $\underline{13}$ 1712
11. Weschler, C. J., Sullivan, J. C. and Deutsch, E., Inorg. Chem. (1974) $\underline{13}$ 2360
12. Gruen, D. M., Gaudioso, S. L., McBeth, R. L., and Lerner, J. L., J. Chem. Phys. (1974) $\underline{60}$ (1) 89-99
13. Green, D. W. and Gruen, D. M., J. Chem. Phys. (1974) $\underline{60}$ (5) 1797-1801
14. Osborne, D. W., Schreiner, F., Flotow, H. E. and Malm, J. G., J. Chem. Phys. (1972) $\underline{57}$ (8) 3401-3408
15. Osborne, D. W., Flotow, H. E., Fried, S. M., and Malm, J. G., J. Chem. Phys. (1974) $\underline{61}$ 1463-1468

7

Hierarchical Minicomputer Support as a Methodological Aid to the Laboratory Investigator

R. L. ASHENHURST

Institute for Computer Research, The University of Chicago, Chicago, Ill. 60637

Over the last decade the use of computers as an integral part of laboratory instrumentation has become routine. Technological developments, however, continue to influence the style of computer-based "laboratory automation." Initially, the connection of a battery of laboratory instruments to a single computer with provision for handling realtime processes on a multiprogramming basis was the norm. Subsequently the availability of the minicomputer, with its ever decreasing cost, and now the introduction of the microcomputer, have engendered more of a trend toward one-on-one computer-to-instrument design. At the same time, the limited capabilities of minimal minicomputer and microcomputer configurations, along with the decreasing cost of digital communications, have led to the idea of connecting individual laboratory computers into backup computer configurations, thereby enhancing the computing services available at the laboratory site.

Although many such "laboratory automation support systems" exist, they differ considerably in their general approach, not to mention their detailed implementation. It is also true that such systems are often not reported adequately in the literature, and even if information is available as to hardware configuration the all-important software configuration features are not made clear.

In this article some aspects of such systems are discussed from the point of view of the laboratory user. The discussion is framed in terms of a particular system, with which the author has been involved, that has been consciously designed from the ground up from a user-oriented perspective. The system having been developed on a university campus for general research use in the physical and biological sciences, it happens that the discipline of chemistry has figured importantly among its early uses. The principles according to which the system is structured, however, are essentially "discipline-free," and it can be argued that this approach presents some advantages even if the system were only to be used in a single disciplinary context, such as chemistry.

A Hierarchical System

Over the past three years the Minicomputer Interfacing Support System (MISS) has been developed at the Institute for Computer Research at the University of Chicago, under a grant from the National Science Foundation.* The MISS project was developed in the context of minicomputers used in the laboratory, although it is adaptable to other uses as well, such as online monitoring of institutional operations (1).

The configuration design is hierarchical, with the laboratory minicomputers at the lowest level (level 0), connected to an intermediate level system dedicated to their support (level 1), which in turn is connected to a general-purpose facility serving these and other needs (level 2). Thus the connections are as shown schematically in Figure 1.

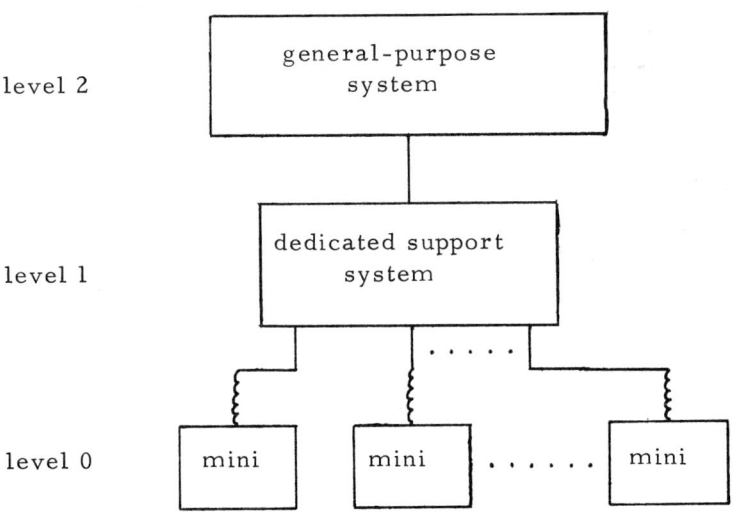

Figure 1.

*NSF Grant No. GJ-33084X

As conceived, the level 0 minicomputers can be of a variety of types and capabilities (at present PDP-11 series and Nova 800 series are supported, with PDP-8 series soon to be added), and the level 2 general-purpose facility could be a computer network instead of a single system (but in fact is only the latter, the 370/168 system operated by the University of Chicago Computation Center). The level 1 system is the heart of the complex, since its operation is the key to the blending of the high- and low-level capabilities offered by the complete system. Although implemented on a minicomputer, it actually is a medium-sized system (a PDP-11/45 with 64K core and card reader, line printer, disk storage and magnetic tape peripherals).

Although some might desire a more detailed system description at this point, this would fall into the common error of preoccupation with systems instead of use. Hence the details will be omitted in favor of a discussion of general considerations. Some more configuration details are given in a recent article (2).

Functions of Levels. Four particular aspects which play an important part in the way MISS is structured are (a) the existence of the intermediate level (rather than having minicomputers connected directly to general-purpose facilities); (b) the concept of the intermediate level providing a "nonresident operating system" for the minicomputers; (c) the provision for connecting a variety of minicomputers, with as little as possible specialized to the individual types; (d) the generality of the connection to the general-purpose system, so that a variety of uses can be entertained.

In brief, the reasons for the importance of these aspects is as follows:

(a) The intermediate level, dedicated as it is to the online support function, affords the possibility of greater reliability and flexibility in that function, as well as provision of more particular services than might be available in the general-purpose facility. Not the least advantage here is a nontechnical but nevertheless very real one, that with the intermediate level system is associated a staff support group organized around the provision of services for online minicomputers. The intermediate level, however, can be viewed in an even broader context--in a very real sense it "mediates" between lowest and highest levels, and enhances both in so doing. For example, it can give the minicomputer at the lowest level access to extended compiling and assembling capabilities carried out at the highest level, and also handle the problems of system access between the levels, making it unnecessary for the minicomputer user to grapple with them. This function is significant even in the present case of a single system at the highest level, the 370/168. If this were extended to access to a general network, such as ARPANET, the mediating function would be even more important.

(b) Rather than functioning merely as a "communications controller" or "front-end" for access to general-purpose facilities, as is true in some hierarchical systems, the intermediate level is specifically designed to support a range of functions that would be characteristic of a medium-sized, and hence in general greatly expanded, minicomputer configuration. Reference to operating system characteristics, then, is made to determine which of them do not need to be resident in the computer which the operating system supports. Although the normal idea of an operating system is that it runs in a computer system to render it more usable, it turns out that a surprisingly large number of its functions can be provided remotely, all or in part. This aspect is specifically discussed in a paper presented at a symposium on computer systems (3).

(c) The design of the hardware and software interface from the lowest level to the intermediate is made general, emphasizing maximum flexibility and minimum cost. This results in a minimal communication/terminal package being designed for each minicomputer variety, which communicates in a standard (logical) way with the intermediate level system. Thereafter an attempt is made to have the system depend as little as possible on what variety of minicomputer it is. Even where this distinction must be made, for example in assembly programs running at the highest level, the manner of invoking the assembler is common to all minicomputers. Hence the effect, as shown by (b) above, is as if each minicomputer uses a standard but medium-sized operating system.

(d) Rather than using one of the more specialized entry modes to the highest level, such as "terminal access" or "remote job entry," the physical connection is general (i.e. channel interface), so that the immediate level system appears as a peripheral processor to the general-purpose facility. This permits various more specialized access modes to be supported by the mediating software in the intermediate level system.

Centralized/Localized Computing. Aspects (c) and (d) above, taken together, represent a crucial objective from the point of view of the laboratory application. Namely, the user connected to MISS from the laboratory can use the minicomputer as a general standalone device, or as a terminal to access the general-purpose facility, both without excessive overhead to what these functions would require in any case. Thus on the spectrum of localized/centralized computing, the user has both extremes available (the standalone localized minicomputer, the terminal-accessed centralized facility). No initial choice need be made to "go centralized" or "go localized," which dichotomy has caused much controversy and expenditure of energy in the political arena of campus computing. By virtue of having both these extremes present in the complete system, the user can get the advantage of a blend of them, that is, adjust to an appropriate point on the localized/centralized spectrum. This is the means by which MISS "enhances" more standard facilities.

The question then arises, what are the reasons for wanting these two extremes in combination? Although it may seen obvious that localized computing is better for some purposes and centralized computing for others, to approach the question of the proper combination requires that an analysis be made.

Laboratory Methodology

For present purposes, experimental methodology may be characterized by functions in five categories, namely: (i) data collection; (ii) automatic control; (iii) human-mediated control; (iv) ancillary analysis; (v) followup activity. The first three of these are the functions which have been so revolutionized by the minicomputer, recording data from instrumentation sensors, and performing online the necessary calculations to do automatically such things as instrument monitoring and adjustment, or to guide an investigator in doing these things during the course of an experiment. The latter two, which may require extensive calculation during the experimental run or afterward, can be carried out on the same minicomputer if it has the computing capacity and the right sort of multiprogramming capability, but generally seem more appropriate for processing by a general-purpose facility, due to volume of calculation and possible output requirements.

One could imagine as a first step at a satisfactory system solution for laboratory needs one where the investigator has in the laboratory a minicomputer for handling (i), (ii), and (iii), and which does double-duty as a terminal device accessing a general-purpose facility for handling (iv) and (v). Capabilities of contemporary minicomputers are certainly such as to permit the kinds of physical interfacing necessary, both to experimental apparatus and to a communications network, and furthermore the minicomputer can readily be set up to perform both roles concurrently, given the appropriate design of interrupt handling procedures (which here would have to be "real-time dependent" for both roles, although more critically so in the first one). There is, however, an additional problem for both groups of functions, that of developing or obtaining the necessary programs to carry out the tasks, and for the latter group the additional problem of handling the input to and output from the centralized facility efficiently from the laboratory. Here is an important aspect of laboratory applications of minicomputers--it must be assumed that programming is a continuing process, as experiments progress or new ones are embarked upon. This is in contrast to the "turnkey" applications in the industrial context, where an application is designed to function indefinitely without reprogramming.

If the combined localized/centralized capability described is also applied to program development, say to provide editing, assembly and compilation on the centralized facility for programs to be tested and run on the localized one, this would seem to round out the needs of the experimenter. But mere provision of capabilities does not render them efficient and effective, and the appropriate structure of the hardware/software configuration combining the localized and centralized facilities is dictated by considerations which must take into account the characteristics of the laboratory application.

Efficiency and Effectiveness. The questions of efficiency and effectiveness have many facets. Although the term "cost-effective" is often heard in this context, where research is concerned there is no monetary measure of the effectiveness, and the researcher is often limited in what can be purchased by somewhat arbitrary considerations such as grant policies and the like. Obviously the "overhead cost" of incorporating a laboratory minicomputer into a hierarchical support system cannot be excessive, or else many researchers will be prevented from using it even if it seems desirable. A prominent part of this cost is in the communications link between the laboratory and the rest of the system. But balanced against this cost and the cost of interfacing is the fact that the hierarchical system may permit a researcher to acquire a smaller minicomputer configuration to obtain equivalent capability to a standalone. This is often the most immediate benefit perceived, and it is therefore important to emphasize that shared peripherals at a site not too far removed from the laboratory are an attractive prospect for the user who has only occasional need for them.

In the MISS design, it is assumed that minicomputer sites can be hardwired to the intermediate facility, which represents a substantial saving over the equivalent capability supported through the telephone system, but of course requires some form of cable network to be available. This concept is well suited to a geographically coherent university campus, and in connection with the MISS project (and in coordination with some other campus projects with similar needs for access to the Computation Center), a basic cable network has been laid on the University of Chicago campus connecting the computer building (which houses both the Institute for Computer Research and the Computation Center) to other selected sites which are close to the locations of possible users. This network consists of twisted pairs, terminated by relatively inexpensive line driving apparatus to handle the desired transmission speed of 9600 baud between the lowest and intermediate levels.

Another possibility allowed for in the design is that of remote concentrators at sites close to several laboratories where MISS users exist. These would be minicomputers with a subset of peripherals, namely those input/output devices for which proximity to the user is important, such as output printers.

User Needs

For the present discussion there may be distinguished three types of user: the investigator, the research assistant and the technician. The investigator is interested in the computer setup only in its function of facilitating laboratory work, and is generally unwilling to expend much effort in absorbing system details, and impatient with what may seem awkward or slow system response. The research assistant can be persuaded to get more involved in system details, if the investigator is convinced that it is really necessary, but is still primarily interested in what the system can do to support the scientific effort. The technician can follow detailed prescriptions but remains laboratory-oriented rather than computer-oriented. These characteristics are purposely made a little dogmatic, and shaded toward the applications side. Although exceptions do exist, where the investigator or the research assistant becomes enthusiastically immersed in system details, perhaps to a fault, clearly system design should aim toward handling the situation where the user needs minimal computer proficiency.

When considering the actual user needs in this way, certain seemingly conflicting desiderata come to the fore. These can be appreciated by considering the words "reliability" and "flexibility" used in two different senses. Reliability can mean dependability for systematic operation according to fixed but possibly inconvenient standards and schedules, as with centra-

7. ASHENHURST *Hierarchical Minicomputer Support* 115

lized systems, or it can mean availability at times most convenient to users and under their control, but without provision of supporting services, as with localized systems. Similarly, flexibility can mean versatility in being able to run a wide variety of routine tasks performed according to general system standards, as with centralized systems, or capability of being modified over short time periods in response to user-specified requirements, as with localized systems.

An attempt has been made in the design of MISS to achieve a blend of these advantages in both senses, and in addition to make them easily accessible to the scientific user by letting the intermediate level serve as facilitator. Some of the system characteristics relevant to this aim will now be described.

Reliability. The separate intermediate level is a significant factor in promoting reliability. The set of functions it provides are fixed, and the demands on them relatively predictable, so that operation can be expected to be stable. The fact that user programs are not run on the intermediate level might be regarded as a missed opportunity by some, but serves to make the system less subject to unexpected load variations, which would also render performance less dependable.

Reliability of the intermediate level disk and magnetic tape storage are particularly important for the laboratory investigator, who must make the decision as to how much of the care of experimentally derived data should be trusted to them. The hardwired access connections, the error monitoring logic built into the line handling software running at the intermediate level, and the design of the operating system at that level to isolate independent functions as separate "processes" all contribute to rendering the instrument-data-to-backup-storage procedure a dependable one. It should be emphasized, however, that the 9600 baud transmission rate from the minicomputer to the PDP-11/45 is not intended to serve the total range of experimental data rates found in practice. For very high data rates, it is appropriate for the minicomputer to be equipped with its own peripheral disk, so that the responsibility for retention of data in massive amounts is not relegated to the shared system but rather to the dedicated one under the immediate control of the experimenter.

Reliability at the highest system level is enhanced by the provision for monitoring, at the intermediate level, the services provided by the general-purpose facility. The purveying of a certain set of "packaged" services appropriate to the minicomputer user (such as compilation and assembly) is regarded as part of the mission of MISS maintenance group, which thereby frees the users from the burden of dealing with the ins and outs of such services "on their own."

Flexibility. In contrast with such "standard" services, which may be made available by the intermediate level in a way such that the user is not even aware of their source, there may be special needs for which an investigator must deal directly with the general-purpose system, and here it is important that the access mechanism not limit such dealing. The previously mentioned aspect (d) of the hierarchical system permits such flexibility to be achieved. In fact, one particular list of development in the MISS project concerns the ways in which access to large job-oriented facilities should be tailored to the particular (nonstandard) case of the user with only a minicomputer and teletype (thus no card reader and line printer).

Finally, the aspect (c) of the hierarchical system cited earlier permits considerable latitude in the way the minicomputer can be operated in the laboratory. Besides permitting it to function either as a standalone experiment controller or as merely a terminal to the general-purpose facility, the interface is designed to allow these modes to be maintained concurrently (by programming so that the minicomputer communication package handles all realtime interrupts). Access to the specialized services of the intermediate level is an added bonus here. But the latter also makes other combined modes possible, such as running program assemblies in the minicomputer while letting the intermediate level disk storage be the repository for source program and object code. This allows the user with the minimal minicomputer configuration to use it for program development without the burden that the lack of peripheral facilities would ordinarily entail, which represents an alternative to delegating the assembly of programs to the general-purpose facility.

Experience. This brief discussion has indicated some of the ways that hierarchical minicomputer support such as that provided by MISS can be made reliable and flexible, without the sacrifice of efficiency and effectiveness. Experience with initial users during the development phase has indicated how much the close interaction between users and developers can facilitate the exploration of these system advantages.

Although the discussion has not been oriented to any particular applications, its relevance to experimental chemistry is at least as great if not greater than to other fields. In fact, the first laboratory use of the system has been in connection with that discipline. The initial external connection of MISS was to the PDP-11 in the molecular beam laboratory of L. Wharton in the University of Chicago Franck Institute. Here MISS not only has provided a particular investigator with useful support even in its development phase, but the user feedback obtained in this way has been invaluable for the project.

At this writing MISS is still in an advanced stage of development as a complete system, although several users have been getting useful work from it for some time. There are currently nine minicomputers connected to it, with plans for fourteen more in the near future. Recent completion of the campus cable network has made MISS accessible from a variety of locations on the University of Chicago campus, and the total of twenty-three minicomputers now or about to be connected are located in a total of five buildings. Applications of these minicomputers cover a range of uses, both laboratory and operational, in the Physical Sciences and Biological Sciences Divisions.

Literature Cited

1. Ashenhurst, R.L., and Vonderohe, R.H., Datamation, (1975) 21 (2) 40-44.
2. Ashenhurst, R.L., Federation Proceedings (1974) 33 (12) 2405-7.
3. Ashenhurst, R.L., in "Computing Systems" (conference proceedings), University of Texas at Austin (1973).

8

Computer Networking at UMR

D. W. BEISTEL, Department of Chemistry
R. A. MOLLENKAMP, Department of Chemical Engineering
H. J. POTTINGER, Department of Computer Science
J. S. deGOOD and J. H. TRACEY, Department of Electrical Engineering
University of Missouri, Rolla, Mo. 65401

The University of Missouri - Rolla (UMR) employs two levels of computer networking in its educational and research activities. The IBM 360/50 of the UMR Computer Center and the other computer facilities of the four-campus, University System are linked via 50 KB data lines to a central IBM 370/168 computer at Columbia. That U-wide network was established to provide the maximum, affordable computer power to every potential user in the University while reducing duplication and its added costs. Figure 1 shows the hardware locations and manner of linkage for the U-wide network, referred to as the Macronetwork by the authors in this paper. In additions to the batch and special jobs processed by Chemistry and Chemical Engineering on the macronetwork, the computer expertise on the UMR campus has drawn together the minicomputers of several academic departments in what we call the UMR Mininet, diagrammed in Figure 2. The two levels of computer networking provide the potential of a vast range of computer services for the chemistry - chemical engineering programs and we will examine present and planned applications.

Teaching and research applications of computers in chemistry and chemical engineering at UMR in the 1960's were accomplished by batch processing at a central, campus facility and by RJE. The Department of Chemistry assembled a software package for spectroscopy (1) that included infrared, ultraviolet and magnetic resonance spectral applications plus an overview of current molecular orbital approximation methods. The programming was developed for student use and provided for the immediate application of theoretical principles to the analysis of complex spectral data.

A Packard model 901A multichannel analyzer was assembled to control repetitive scan, NMR experiments and acquire data in CAT and kinetic studies. The configuration, as shown in Figure 3, allowed for indirect interfacing to the IBM 360/50 via nine-track magnetic tape when complex intensity or frequency analysis was

8. BEISTEL ET AL. *Networking at UMR*

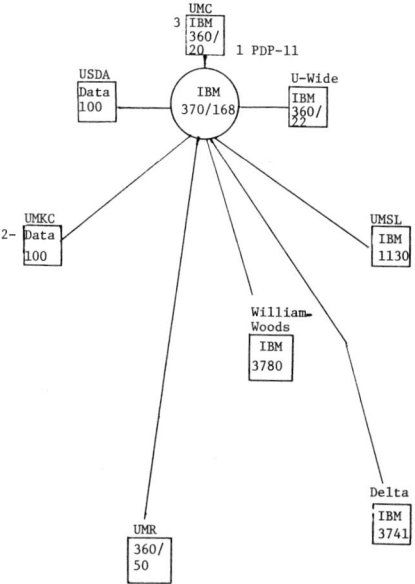

Figure 1.   U-wide computer network configuration ca. 1975

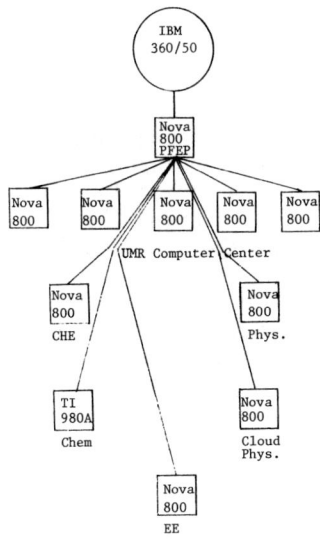

Figure 2.   UMR Mininet

required. Software was developed to accommodate those applications using modifications of LAOCOON III (2), DECOMP and ASSIGN (3) and other programming developed in house.

Further computer applications to chemistry included paper tape output from a scintillation counter and readout from the Reynolds 60 degree sector mass spectrometer, along with curve fitting-plotting routines for the physical chemistry and instrumental methods laboratories. The programming was resident on disk at the UMR Computer Center and initiated as a batch job. Remote job entry via terminal was limited to less than one per cent of the computer activity of the Department.

In the Chemical Engineering Department, batch computer applications included a wide range of simulation and data analysis projects. The undergraduate classes in process design made extensive use of design programs such as CHESS, and considerable work was done to expand program options and capabilities. Distillation projects included development of a simulation program applicable to multi-component, sidestream columns. Control studies involved dynamic simulation of processes and control systems for design of optimal and sampled-data control strategies.

Analysis of experimental data was of prime importance to research on mixing and on enzyme reaction kinetics. The mixing studies included the modeling of laser-doppler anemometer data to predict mixing patterns in process vessels. That research generates a large amount of data in looking for time variations and very high frequency fluctuations. Enzyme reaction kinetics studies have involved the modeling of reaction rates with a mass spectrophotometer. Some of the reactions studied are quite slow and required eight to twelve hours for a single experimental run. Computer analysis of data required manual digitizing, card punching and batch processing using the IBM Scientific Subroutine package as well as user developed programs for regression analysis.

In 1973 all batch processing control was transferred to the macronetwork and only plotting and specialized jobs were spooled to the IBM 360/50. Remote terminals were linked directly to the IBM 370 system under TSO. During this period the computer graphics capabilities of the UMR Computer Center and the Department of Electrical Engineering were expanded by acquisition of Data General Nova 800 units and the two facilities were linked by phone line. The slow data rates (110 B) of the dial-up line were unacceptable to the users in Electrical Engineering, however. Because several stand alone Nova 800 computers were in operation in other departments on campus, the long-range potential of a mininetwork was evaluated and relatively high speed (19.2 KB) data lines were installed to permit its development. Link up was accomplished by the staffs of the computer center and electrical engineering and a number

of innovative programs were initiated. (4) In 1974 the
Departments of Chemistry and Chemical Engineering purchased major
computer hardware, extending the mininetwork to its present
configuration.
 The linkup of the Texas Instruments 980A computer of the
Department of Chemistry provided a number of challenges. First,
the computer was purchased from JEOL to support a JEOL JMS D-100
high resolution mass spectrometer. The programming system for
its dedicated function was developed by JEOL in a coded format
that defied ready interpretation and it required casette input
via Texas Instruments ASR 733 device. It was obvious to us that
no stand-alone computer applications could be developed before
the mass spectral facility was converted to a disk operating
system. Several software approaches were taken before the
Texas Instruments bootstrap program, MHBOOT, was used to read the
mass spectral programming into core as data from disk, avoiding
the five minute load time from casette.
 Another reason for that approach may be less obvious to a
potential user of a dedicated system. The operator of the mass
spectrometer has acquired data from the Varian 2700 gas
chromatograph and the JMS-D100 mass spectrometer for as many as
two hundred, 4K word scans. He must normalize the mass spectral
data and choose his output device and formats before his study
is completed, a time-consuming process. After the data are
stored on disk the operator can return to the analyses at his
convenience and is limited only by the available disk storage
of his unit.
 The TI980A computer system as supplied by JEOL had a Calcomp
565 plotter. The plotter was not a Texas Instruments - optional
accessory at the time of purchase, so its use in the stand alone
operation required some software development. Using notes
supplied by JEOL and output from the IBM 360 plotter package an
assembly language routine was written. It provides the same
quality of plots presently offered by the IBM 360/50 system at
the UMR Computer Center. The plotter is used in support of
linear least squares programming at the present time. Its
primary support function remains in mass spectral output.
 To facilitate mass spectral file storage and searching as
well as planned, off-campus communication, a communications
package was developed to provide a viable link to the Mininet.
We found that the communications module available from Texas
Instruments was not capable of multilinking, although it is a
versatile module because of its software support. The staff at
Washington University was kind in providing the circuit diagram
for an adaptation of the TI module, but on examining their
schematic it was concluded that significant improvements could
be made at reduced cost. The full schematic on our communications module will be published at a later date, but meanwhile
a working configuration is shown in Figure 4. The transmitter-
receiver functions provide communications over two one-half mile

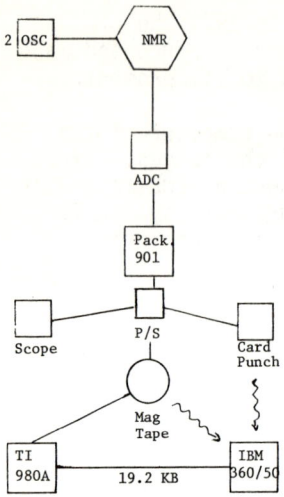

Figure 3. Repetitive scanning magnetic resonance facility

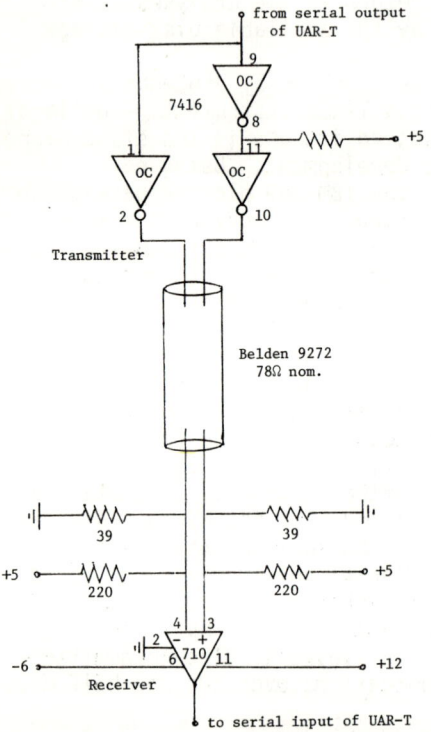

Figure 4. Differential transmission scheme used in Mininet at UMR

data lines. This new communications module has the advantage of multi-data rate and multi-data line hookup at one I/O board location on the computer. We actually built two modules — one for external communications like the Mininet and one for internal, remote sites within the Chemistry - Chemical Engineering building.

One unit is dedicated to the Mininet at present. The planned development of the second communications module is shown in Figure 5. The first stage link to the Packard 901 multichannel analyzer required no additional interfacing at the analyzer, because the analyzer is equipped with a Packard 970 parallel-to-serial converter with front panel switching of I/O options. For this link only the cable connectors for the high speed paper tape option were needed and a 1000 character-per-second data rate was employed for I/O.

The present I/O from a ^1H or ^{19}F magnetic resonance experiment proceeds as follows: The analyzer system acquires magnetic resonance data, storing each scan on an assigned magnetic tape record. The data are then transferred via a data line to TI980A disk, transferred via the ininet to the IBM 360/50, which in turn transfers the data to the IBM 370 system via the macronetwork for processing. The output from the IBM 370 - **LAOCOON III** program is transmitted back to the multichannel analyzer via a reversed sequence and is plotted on the A56/60 recorder under the spectrogram. Because we do not operate the TI980A computer under an executive program, we must gain its attention and monitor the initial transfer of data. Entry to the Mininet also requires a telephone alert at the present time.

While the full transfer operation seems cumbersome, in practice the steps are simple and convenient. Further, some processing can be done in house at the option of the user. A complete spectral analysis requires 380K words of core but only about 4 to 6K words of core are needed for most kinetic studies. In fact some intermediate stages such as integration can be accomplished on the multichannel analyzer if the base line is linear. If not, base line adjustment is accomplished using the program, **BLINE**.

We plan to locate a remote ASR 733 module in the physical chemistry laboratory for use in a short course in laboratory computing. Presently a Wang 362 calculator serves that purpose but does not accommodate the range of calculations required during the two-semester course, forcing the class to use the TSO links to the IBM 370 at Columbia. We also plan to acquire a computer graphics unit for attachment to the TI980A, which will be used to introduce molecular orbital concepts in the physical chemistry laboratory. Research applications are planned also.

In January 1973, the Chemical Engineering Department purchased a Data General Nova 800 computer system to support data acquisition and control activities within the department.

The computer and peripherals were bought as a system with Data
General taking the responsibility for integrating components
from a number of manufacturers. The system included a full
complement of computer and process I/O equipment and a fixed
head disk for mass storage. Subsequent purchases, including a
cartridge disk drive, additional core, and a CRT terminal,
resulted in a very flexible system (Figure 6).
   The Nova 800 computer selected is a 16-bit word machine
organized around four accumulators. Options included are
hardware multiply and divide, real time clock, power monitor/auto
restart, and automatic program load. The original 16K words of
memory has since been expanded to 24K.
   Equipment purchased for computer I/O consists of an ASR-33
teletype, punched card reader, and 30 character per second,
portable terminal. An alphanumeric CRT terminal has since
been added and is now used as the primary computer console.
   Process I/O equipment was selected to provide flexibility
in interfacing with a variety of laboratory instrumentation. A
wide-range analog-to-digital (A/D) system can accept up to 16
differential inputs with full scale ranges of ± 2.54 mv to
± 10.24 v in 13 programmable steps at a rate of 200 samples per
second. A high level, 4 channel A/D system can take up to
50,000 samples per second and deposit the data directly into
computer memory through a data channel. Additional process I/O
consists of 6 digital to analog (D/A) channels, 16 digital
inputs and 16 digital outputs.
   Present mass storage consists of a rapid access, fixed head
disk and a removable cartridge disk. With 256 K words of
storage, the fixed head disk provides fast, temporary storage of
data and programs. The several 1.25 million word, removable
disk cartridges give users permanent storage space for programs,
data, and computer disk operating systems.
   Vendor supplied software for the Nova includes a real time
disk operating system and real time Fortran. This software,
along with utility programs for editing and file maintenance,
provided a starting point for development of programs for data
acquisition and control.
   Fortran callable, assembly language programs were written
at UMR to support A/D, D/A, and digital I/O. Therefore, the
majority of users can write application programs in Fortran with
no need to learn assembly language programming.
   Perhaps the greatest incentive for purchasing the Nova
system for Chemical Engineering was the existence of several
Novas in other departments at UMR. The commonality of equipment
and system software allows two people to handle all maintenance
and modification of hardware and development of system software.
Interfaces designed for one Nova are immediately applicable for
all Novas at UMR. These have been important factors in the
development of the Mininet.

8. BEISTEL ET AL. *Networking at UMR*

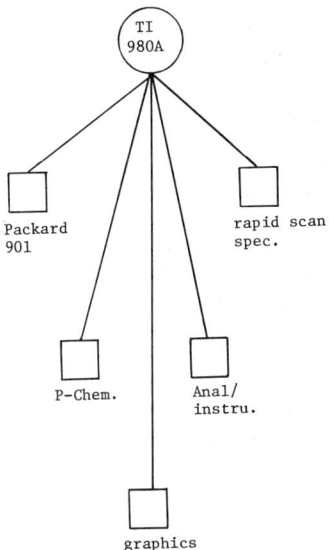

Figure 5. Projected development of TI980A local network

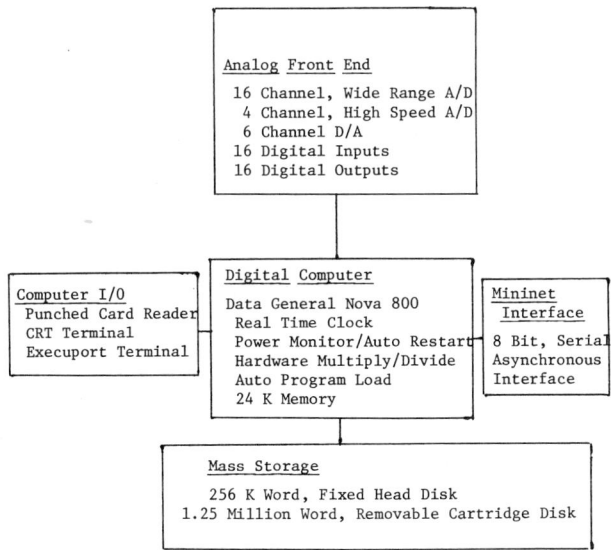

Figure 6. Chemical Engineering Computer System

Initial applications of the Chemical Engineering Nova have been as a stand alone system. Generally, one user at a time will use the system for data acquisition and control. However, simple data acquisition tasks may be executed in a foreground partition while program development or complicated control programs are conducted in background. The following indicates present applications of the Nova — both as a stand-alone system and as a part of the mininet.

The Nova system has been applied to data acquisition in mixing studies. A 0-10 volt signal output from the laser-doppler anemometer is sampled by the high speed A/D system at a rate of 50,000 samples per second. Data is stored directly into memory until space is exhausted. During the data acquisition, the computer begins transferring data to the fixed head disk. After all the data stored in memory has been transferred to disk the sampling process continues. In approximately five seconds space is exhausted on the fixed head disk and the experimental run is complete. Disk data is retained for processing on the Nova as a batch job.

Using the Mininet, this data can be transferred directly from memory to the IBM 360/50 for high level analysis and plotting of results. The interface with the mininet is a 8 bit serial, asynchronous, full duplex interface. The interface can transmit data at four different rates — 9.6, 19.2, 38.4, and 76.8 K band — under program control.

In the area of digital computer process control, the Nova has been used to automatically control two experimental processes in the process control laboratory. One process is a simple liquid level-flow system (Figure 7) used to demonstrate feedback control principles and evaluate digital control algorithms. A second process mixes hot and cold water to produce a combined stream of controlled temperature and flow rate. This apparatus is used to demonstrate cascade, ratio, and feedforward control and to design and evaluate non-interacting, multivariable control strategies.

These sets of equipment use conventional industrial type sensors, control valves, and transducers to provide computer compatible signals. Input signals are accepted into the computer via the wide-range A/D system. The D/A system provides control signals to valves while computer set digital outputs are used to start and stop pumps.

These process control applications utilize multiple task programs for acquisition, control, alarming, startup, and emergency shutdown. Up till now, this work has been developmental so that during operation the user requires editing and compiling capabilities.

Future applications in control will utilize the mininet for periodic optimization or evaluation of complicated computer models of lab processes. Plans are to link continuous

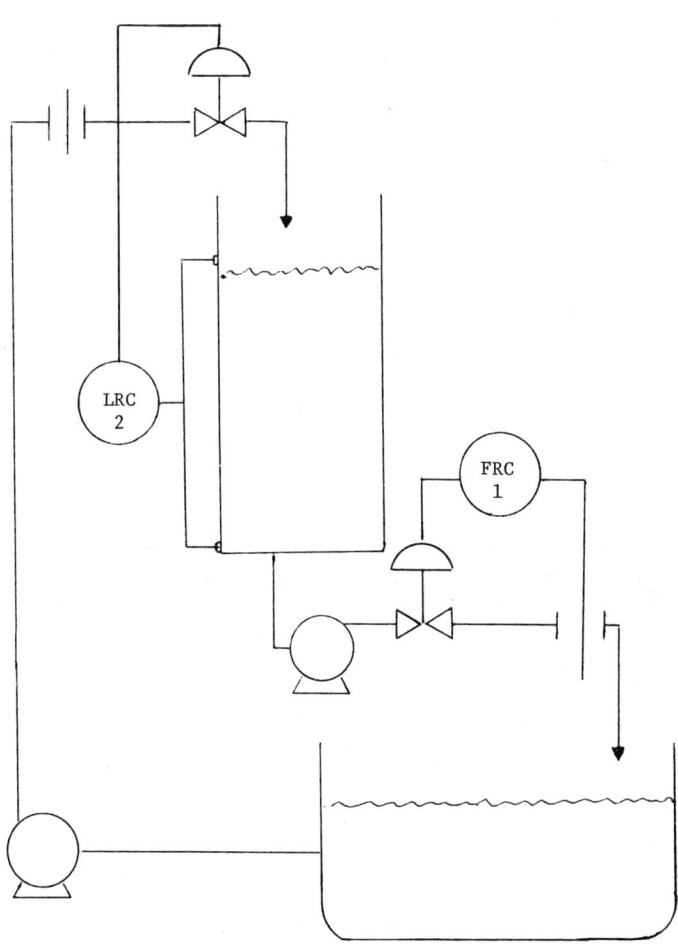

Figure 7. Liquid level-flow process used for computer control studies

distillation equipment in the unit operations lab with the Nova for data acquisition and control. Large simulation programs on the IBM 360/50 will use the lab data as a comparison for evaluation of steady-state and dynamic characteristics.

In kinetic studies of enzyme catalysis, data may be acquired by the computer during lengthy experimental runs. A 0-10 volt output from the mass spectrophotometer is sampled at relatively slow rates and for long periods of time. To allow continued program development during these times, the sampling is done by a foreground program, stored on disk, and analyzed at a later time.

Future applications in this area will involve much faster reactions. Data will be sent to the IBM 360/50 via the mininet for regression analysis. The results will be returned to a terminal in the kinetics lab to guide subsequent experimental runs.

Abstract

The University of Missouri has two levels of computer networking in operation, a macronetwork to the central IBM 370 system at UMC and subnetworks at UMC and UMR. The subnetwork at UMR is a mininetwork of seven Data General Nova 800's and one Texas Instruments 980A, tied by direct 19.2 KB data lines to a Data General Nova 800 link to the IBM 360/50. The TI980A (Chemistry) facility can be dedicated to on-line data acquisition with a JEOl D-100 mass spectrometer or act as a stand-alone terminal with plotter. The Data General Nova 800 (Chemical Engineering) is a stand-alone unit equipped for a variety of multiplexed, data acquisition applications. Hardware and software development for the Nova 800 is in advanced stages because of advanced applications such as computer graphics in other departments, while the TI980A link requires a variety of innovative developments to act as a functional link in a Nova network. The details of hardware and software development are discussed in the context of applications to chemical problems.

Literature Cited

1. Beistel, D. W., J. Chem. Ed., (1973), 50, 145.
2. Bothner-By, A. A., and Castellano, S., "LAOCN3", Mellon Institute, Pittsburgh.
3. Lusebrink, T. R., Ph.D. Thesis, University of California at Berkeley, August 1965.
4. Tracey, J. H., and Pottinger, H. J., Proc. IEEE, (August 1975), in press.

9

Computer Assembled Testing in a Large Network: The SOCRATES System

WILLIAM V. WILLIS

Chemistry Department, California State University, Fullerton, Calif. 92634

OLIVER J. SEELY, JR.

Chemistry Department, California State College, Dominguez Hills, Calif. 90747

    The use of computers in the formal educational process is increasing in diversity and extent. As scientists, chemists have often used computer techniques to solve a wide variety of problems. As educators, we are beginning to use computers as a resource and tool to develop more effectively and efficiently the knowledge and skills of our students. Much of this use has been by individual instructors, and therefore has been developed with highly specialized goals in mind for a small number of users. For example, on-campus CAI type instruction in the analysis of NMR spectra, or solving stoichiometry problems etc. This paper describes the formulation and implementation of a large general purpose program called SOCRATES (Student Oriented Classroom Analysis and Test Evaluation System) designed for operation in a 19 campus computer network serving 290,000 students and 16,000 faculty in the California State University and Colleges. SOCRATES is a data management system designed to produce and process exercises (tests, tutorials, homework, etc.) as directed from collections of questions, or data banks, in thirteen fields: chemistry, physics, mathematics, psychology, economics, US history, accounting, FCC review, biology, testing and measurement, counseling and guidance, political science and data processing.

    The concept of generating tests by selecting prepared questions from a collection is certainly not novel. Instructors have traditionally maintained in personal collections file cards which are searched and given to a typist for suitable duplication. On a larger scale, commerical testing services have maintained computerized files for approximately twenty years, and presently more than one hundred educational institutions in the USA and Canada have

collections stored on magnetic tape. Usage of the
latter has not been great for a variety of reasons,
including poor communications, limited accessibility,
and teacher attitudes. One SOCRATES critic summed up
these attitudes in the remark "You are just automating what we all do." We will attempt in the following
paragraphs to describe this very automation, and the
reactions of educators, students, and others which
lead us to believe that it will produce a profound
change in certain aspects of the educational process;
indeed, the large scale application of computer technology to education thru data banks and networking
is feasible, economical, and pedagogically sound.

The SOCRATES Retrieval System

Our data banks were developed initially for use
in a retrievalsystem described by Lippey, et al. (1).
It was impossible for all campuses in the network to
access the data banks directly due to hardware and
management constraints. The system was redesigned and
rewritten in ANSI COBOL to operation on a CDC 3300
central computer. Formal institutional support for
the system has been available for the past two years,
during which time the chemistry data bank has grown
from 2,000 to 10,000 items.
The retrieval system allows the user to select
items from the bank employing a variety of parameters,
either singly on in concert, to execute the search.
These parameters include a category number (see
below), level of difficulty, a behavior level (requiring either the demonstration of knowledge or the
application of knowledge), key words, item source
identification, and a specialized search parameter
whose function varies from bank to bank. Two other
general search parameters are available which allow
blocks of interrelated questions (macro items) and
items which require materials not in the data bank,
such as slides, audio or TV cassettes and figures,
(enhanced items) to be selected or suppressed from
selection. Each exercise can contain up to 150 items,
and may be edited by the user to delete or add items.
Up to nine versions of the same exercise may be requested with different orders of presentation of
the contents, and the exercise can be produced by
line printer on either console paper or continuous
form reproduction masters at the users request. Objective items in the exercises (multiple choice, true-
false) can be computer scored; individual student
performance may be then compared with records of

averages for past responses maintained by SOCRATES. The SOCRATES system can be used either in a batch mode or a timesharing mode. No knowledge of computer programming is necessary to use the system's educational features. The content, generation, administration, and grading of SOCRATES exercises is controlled by the user. Typically, a request submitted by telephone will result in the overnight production of the exercise, and delivery the following day.

The most important feature in the design for user access to each data bank is the subject matter classification system. The SOCRATES system uses ten major categories which are arrayed in a heiarchy for searching. Table I shows the main subject classifications for the chemistry bank.

Table I
Major Categories in the SOCRATES Chemistry Bank
00000	I.	Reserved (non-standard items)
10000	II.	Introductory Chemistry
20000	III.	Atomic Theory and Structures
30000	IV.	Chemical Periodicity
40000	V.	Stoichiometry and The Mole Concept
50000	VI.	Kinetic and Equilibrium
60000	VII.	Solution Chemistry
70000	VIII.	Bonding
80000	IX.	Thermodynamics
90000	X.	Topics

Within each major category, sub categories are established, as shown in Table II which shows a portion of the detailed classification system for solution chemistry.

Table II
Partial SOCRATES Category Listing for Solution Chemistry
60000	VII.	Solution Chemistry
.....	.....	
62000		B. Acids, Bases, and Salts
62100		1. General Properties
.....	.....	
62140		(D) Acids and Bases
62141		(1) Strong Acids (including oxidizing properties)
62142		(2) Weak Acids
.....	.....	

The exercise generation program would, upon request for solution chemistry items, search all items between the limits 60000 and 69999; if instructed to search for items dealing with acids, bases, and salts, categories 62000 thru 62999 would be searched, and so on. In this manner, a 10,000 item bank can be searched to produce a 100 item test in 1.07 minutes total time, including I/O, at a cost of $3.46. The user is provided a copy of the test with all questions sequentially numbered, a cover identification page which also has a tell tale to insure privacy of the exercise, a test request page showing the request submitted, and an item statistics page which lists correct answers, item sources and other information.

The content and classification system of the data banks is controlled by the data bank coordinators and the SOCRATES director. Their personal viewpoints will be reflected to some degree in the classification system. The Chemistry bank follows a generalized pattern derived from the format of introductory texts, since the principal use of the bank is envisioned to be for introductory courses with large enrollments. As an example of another approach the U.S. history bank is arranged chronologically. Irrespective of the classification used within the required heirarchy, all banks are being developed so as to permit the greatest flexibility and promote user acceptance. Thus items which favor particular pedgogical techniques, such as programmed learning, are discouraged.

Test generation is but one function of SOCRATES Figure 1 shows a block diagram of the control points and software modules. The batch test generator has been described above. The interactive test generator (ITG) and scoring modules can operate in two modes. The first mode allows the teacher to devise an exercise interactively from a remote console using the same search parameters as provided in the batch mode.

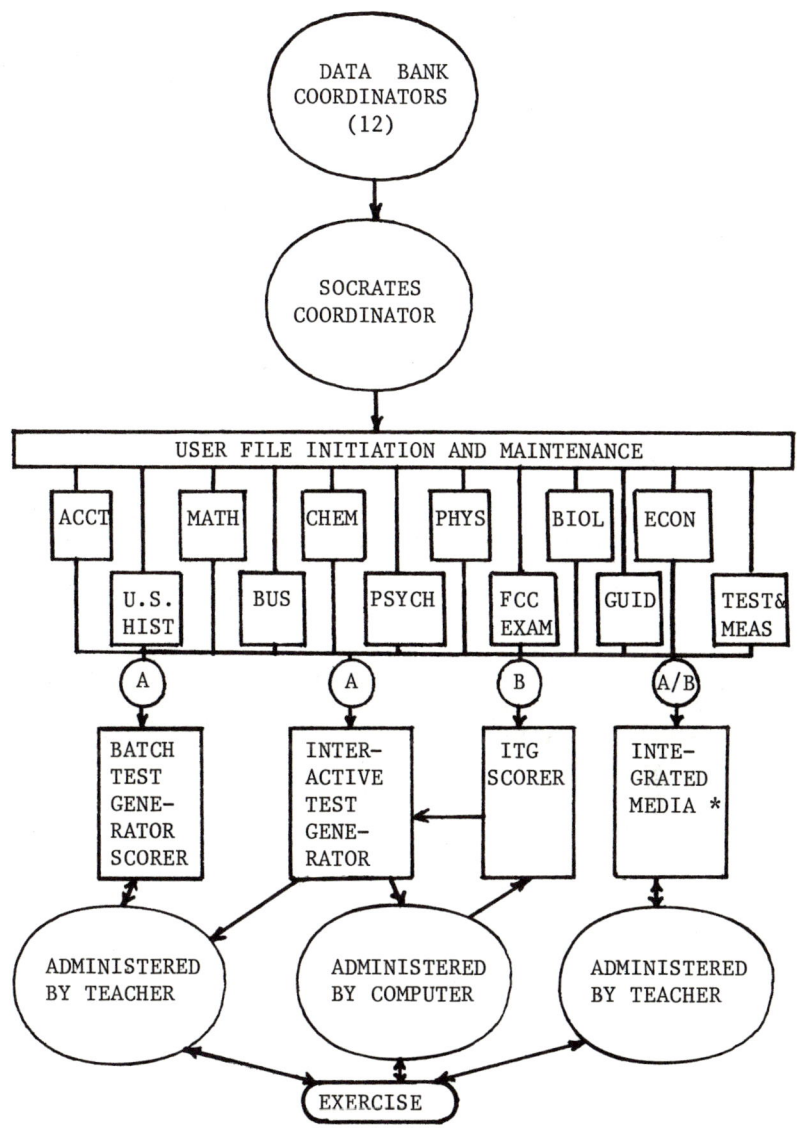

A = Content and scoring determined by teacher.
B = Content and scoring determined by student.
* = Module under development.

Figure 1. Principal SOCRATES modules and control points

After editing output can be provided at the job
initiation site (hard copy), or by line printer from
the Los Angeles data processing center. The second
mode allows the student to choose items from a group
selected by the teacher, and to answer them for credit.
The student has options as to the difficulty (hence
point value) and the content (the question presented
may be refused) of the exercise. The ITG module does
not provide CAI or programmed learning in the sense
that solution hints are not provided, and the content
of subsequent questions is not necessarily contingent
on student response. ITG scoring is provided by a
separate module.

The integrated media module (IM) will be developed to use types of enhanced items in the data bank
which require the use of information or devices
maintained outside the SOCRATES data bank, and therefore not directly under control of the software.

Presently all enhanced items in each bank are
keyed to resource figure files, copies of which are
provided for each SOCRATES user on request. These
are typically line drawings specifically prepared
for ease of duplication by Xerox or spirit process
masters. The IM module would allow users to include
questions based on local resources such as TV tapes,
slides, film loops, etc, this is, materials which
would not normally be duplicated and given to students
individually in order for them to complete the exercise (e.g., periodic tables).

Computing Hardware in the California State University and Colleges CSUC

The hardware currently in use in the CSUC system
is shown in Table III. Networking in the usualy sense
is limited for data processing to links between local
computers and the State University Data Center (SUDC)
in Los Angeles.

Table III. Computer Hardware in the California State University and Colleges System

Campus	Enrollment (Fall, 1974)	Time Sharing Ports*	CDC	Other
Long Beach	31,228	11	3150	
San Diego	30,564	12		IBM 360/40
San Jose	26,829	7	3300	
Northridge	25,377	11	3170	
Los Angeles	23,679	7	3150	
San Francisco	20,855	12	3150	
Fullerton	20,053	7	3150	
Sacramento	19,662	8	3150	
Fresno	15,331	8	3150	
San Luis Obispo	14,434	7		IBM 360/40
Chico	12,680	6	3150	
Hayward	11,711	8	3150	
Pomona	11,099	6	3150	
Humbolt	7,290	3	3150	
Sonoma	5,798	2		NCR 200
Dominguez Hills	5,747	2		Honeywell 20/20
San Bernardino	3,501	2		Honeywell 20/20
Bakersfield	2,897	1		Honeywell 20/20
Stanislaus	2,800	2		Honeywell 20/20
State University Data Center			3300, 3300	
Totals	291,916	112	14	7

*All timesharing operations use the CDC 3170 memory and central processor at the Northridge Campus; ports may have teletype, CRT, or both types of terminals; access is controlled by the SUDC CDC 3300 executive.

Data files in other computers may be accessed indirectly through the executive CDC 3300 at SUDC by prearrangement. The demand for memory or computational speed to date has been met by this combination, so that direct link between campus computers has not been needed. However, the demand for timesharing operations has far exceeded capacity, partially because of the low transmission rates (110 baud) dictated by the existing equipment which must span some 1200 km (maximum) between terminals. To meet this need, Hewlett Packard 2000 computers will be installed locally for hard wired timesharing at higher data transmission rates. The impact of these factors on SOCRATES use and networking will be discussed in a following section.

The Chemistry Data Bank

The chemistry data bank was first compiled in 1972-73 using items donated by the Educational Testing Service and the Monmouth, Oregon City Schools. The modification of these existing questions allowed us to quickly develop the bank to a useful size. The project was supported by an Innovative Funds Grant from the Office of the Chancellor, California State University and Colleges. The initial intent was to develop items covering the first year of college level chemistry. At the project termination date, the bank contained 4000 separate questions, primarily of the machine-gradable multiple choice type. The manpower required for production was as follows: 1/2 manyear for classifying and proofreading keypunched items; 1/2 manyear of keypunch time; 1/2 man year for coordination of these efforts and data processing. (The SOCRATES software system itself has required some 5 manyears for development todate). In the past year the bank has been expanded to 10,000 items. Items have been donated by Eastern Michigan University, Michigan State University, and the University of Pittsburg. This expansion has required approximately 3 manyears of effort.
    The bank is now being used by five percent of the registered users each day to produce quizzes, homework assignments, tutorials, diagnostic tests, challenge examinations, and final examinations. Other bank uses (over twenty have been reported) include production of copies of the entire bank on paper on microfiche for student review in libraries or learning center, and production of blocks of "unsuitable" questions which are given to students to

critique and improve. A popular feature is the production of a general chemistry placement examination which is regularly administered to all first time chemistry students at four campuses. This has greatly facilitated comparisons of student populations, and has led to a study of the potential of the placement test as a counseling tool for incoming freshmen.

One chronic problem created by the diversity of hardware within the system and the language requirements of chemistry, physics, and mathematics has revolved around use of standard business print chains on line printers. Many commonly used scienitic symbols are not available, most importantly sub - and superscripts and lower case letters. We have developed a set of conversions which is serviceable, and meets the following criteria: (1) easy to learn; (2) uses standard business print chains; (3) readily interconvertible to yield similar output by local computer software; (4) will require minimal reformating as appropriate print chains are acquired. The conventions which have been used are perhaps disconcerting at first glance to the user who has a long-standing familiarity with commonly accepted chemical representations, but students readily learn these conventions. A few highlights follow as examples. Elemental symbols (two-letter) are underlined: thus cobalt is written as $\underline{CO}$ which distinguishes it from carbon monoxide. Subscripts are shifted a line down, and superscripts a line up: $H_2^{238} 50$; $\underline{TH}_4$. Equilibrium is represented by $\Longleftrightarrow$, resonance by $\longleftrightarrow$, and reaction by $\longrightarrow$. Questions using electron dot formulas or Kekule structures are bete' noire of the standard print chain character set, and have generaly been supplied separately as enhanced items with prepared figures.

<u>Usage of the Chemistry Data Bank</u>

Most test requests are made via voice telephone directly to SUDC, with a smaller percentage initiated by remote batch. Since a typical exercise may contain 1000 lines, most users elect these options rather than interactive test generation which must proceed at 110 baud. The actual amount of teacher time involved is perhaps slightly less than that required by traditional methods, but the following advantages a^crueby SOCRATES usage: (1) the test may be

conveniently edited any number of times; (2) the test
is produced neat and error-free on reproduction
masters ready for processing (thus reducing the amount
of proofreading and typing necessary); (3) the test
is machine graded; (4) statistics are provided on
student performance. The net result is that teachers
spend their time in two of the tasks most deserving
of their attention and expertise, namely selecting
the content of the exercise and evaluating student
performance. This should provide more time for
the teacher to spend in scholarly and academic
endeavor, and otherwise benefit the student.

However, acceptance of SOCRATES, if demonstrated
by increasing usage, is slow to come. Obviously
professors do not have the same degree of familiarity
with educational computer usage as with research
oriented usage, or with computer produced output
as with materials produced by secretarial staff.
Complete, hands-on, immediate control of all phases
of exercise generation by the individual is a
strong habit. How deeply this process is ingrained
in us is amply demonstrated by some faculty who
request tests, perhaps edit once to add and/or
delete questions, obtain the final output on reproduction masters, then have the secretary retype
the entire exercise with minor wording changes. Others
prefer to examine a listing of the entire bank contents, select items, and have a secretary type the
exercise directly from the listing.

Our experience has shown that science teachers,
as opposed to liberal arts teachers, are more
receptive to the idea of "computerized" testing. This
is due to two factors: the scientist of times has a
closer acquaintence with the capabilities, modes,
and limits of computer usage; the nature of scientific study which lends itself more readily to a
quantitative approach to the subject, which facilitates
development and use of item banks suitable for
computer administration. The challenge therefore
lies in developing questions which are both
pedagogically effective and computer managable (e.g.,
machine gradable).

We have encountered a wide spectrum of
reactions to SOCRATES machine gradable question
format among educators. Many testing and counselling
officers and community college teachers are extremely
enthusiastic. Practically all of the chemists we
have talked to feel that our item bank, if not the
entire SOCRATES system, is a valuable resource for
training their students. Often times the most

enthusiastic supporters are those teachers who have
already developed their own personal item banks.
Although occasional objections are raised to question
phrasing or printer conventions, people actively using the banks items have found their students have
little difficulty in adapting to computer produced
materials.

A popular feature offered by SOCRATES is the
multiple version option. Up to nine versions of the
same exercise can be generated, edited, and scored
as easily as one. The large size of the chemistry
bank permits multiple exams to be developed on the
same subject matter for use by course sections meeting
at different times.

One criticism of SOCRATES is raised at user
workshops and informational meetings which deserves
some comment, as it also reflects habituation to the
traditional testing process as well as a philosophical
viewpoint. Concern is expressed about the effectiveness of objective multiple choice questions as
teaching and evaluation tools in subject areas such
as chemistry where great emphasis is placed on
developing the students abstract reasoing and conceptualizing skills. Many of us have very definite
goals for the content, philosophy, and appropriate
language for questions. These frequently involve a
demonstration by the student of some logical set
of operations which relate to the scientific principle
we are asking about. The ability of multiple choice
questions to meet these goals depends on their
design, rather than their inherent approach. By
careful choice of options, incorrect responses can be
made to pinpoint specific errors in students' reasoning or approach to a problem, at any level. Thus an
examination of the item statistics (student responses)
to an exercise can tell not only general student
performance, but identify individual deficiences as
well. This is not to say that multiple choice questions will supplant traditional subjective items such
as essays, etc. Indeed, the reserved category in
the classification system (00000) includes these.
However, our experience is that they are an effective
teaching tool, as demonstrated by the fact that
students from SOCRATES aided prerequiste classes
perform as well in subsequent classes as other
students.

We are not discouraged by the slow acceptance of
the utility of item banks. They hold great promise
for allowing us to do what we already do by automating certain non crucial steps. As teachers we

can thus reach more students, more rapidly, more
inexpensively, with fewer errors, and at many various
levels. We view SOCRATES as a tool to allow us to
return to the individual student and his problems in
this day of mass education.

The Future of SOCRATES-Type Programs in Networks

Since SOCRATES is basically a data management
system using a moderately sized data base, the main
advantages of networking lie in providing rapid
service to a large geographical area. This could be
extended globally via satellite using the EDUCOM,
CYBERNET, TYMSHARE, or similar facilities. Phone
links have already been used to demonstrate SOCRATES
in Pennsylvania, Illinois, and Texas; satellite
relay demonstrations are planned in France and Mexico.
The question therefore is not one of technology, but
of user demand. The current national trend is for
parallel independent development of item banks. We
have honored a number of requests for console paper
or magnetic tape listings of the chemistry bank,
and encourage the free exchange of items. However,
such exchanges are at times troublesome due to
interschool hardware and/or software disparities. The
problem of standardization to facilitate trans-
ferability has been recognized and is being addressed
by groups such as CONDUIT. Suffice it to say here that
good planning and communication is needed to prepare
independent banks and software for an easy merge as
the need arises. In the immediate future total
duplication and transfer of items banks will be the
rule , since most educational institutions prefer to
support their own independent operations.

The need for networking to facilitate timesharing
operations such as the SOCRATES ITG module, or
general computer assisted instruction (CAI) is questionable for the near future, primarily due to
hardware costs of providing ports and terminals for
even modest class sizes. Typical installations in
the CSUC system have one to three terminals at each
port. Ports are located in accessible places so that
students may use timesharing during their free periods,
and instructional use is planned on that basis. Since
system timesharing use is heavy (15,000 hr/month
average connect time for the Fall 1974 semester,
22,000 hr/month peak) minicomputers are being installed on each campus for localized timesharing. Language
and active storage resources will be limited, so
SOCRATES cannot be locally implemented. It is felt

that all SOCRATES banks could be implemented using the disc space released after transfer of certain operations to the local minicomputers.

We envision that SOCRATES operations will evolve primarily using RJE (remote job entry via local computer to the SUDC CDC 3300 executive) and either local line printers for remote areas, or a central line printer at SUDC and twice daily courier service for the Los Angeles service area. As with any evolving system, new uses will be added as the demand for them arises. Management features such as preparing class grade lists and some format options have already been discussed. Users may elect to write their own programs using the item banks as a resource, but the lessons learned from PLATO experience should be obvious: from 50 to 200 hours of programming effort are required to produce one hour of interactive program.

Perhaps an observation by Marshall McLuhan will also appropriately reflect the coming state of both the SOCRATES effort, and educational computer networking for information retrieval: "In the age of Xerox, every man is a publisher". With the advent of large accessible data bases, teachers can readily publish a wide variety of truly educational aids according to the dictates of their personal standards.

Literature Cited

1. Lippey, G., Toggenburger, F., and Brown, C.D., Assoc. for Educ. Data Syst. Journal (1971), March, p. 75.

10

The Impact of a Computer Network on College Chemistry Departments—The Iowa Regional Network

WARREN T. ZEMKE

Department of Chemistry, Wartburg College, Waverly, Ia. 50677

On the surface, the two main purposes for the existence of a regional computer network are (i) to provide large-scale computing capabilities for medium-sized institutions and (ii) to share in developing the use of instructional computing (1). The purpose of this paper is to examine the impact of these network goals from the perspective of the user (the remote institution or the "fingers of the hand") rather than the provider (the central computer facility or "palm of the hand"). Attention will be focused on the perspective of Departments of Chemistry at twelve of the thirteen remote institutions making up the Iowa Regional Network (officially entitled the Regional Computer Center, or RCC). There are numerous literature sources available for a broad, non-academic discipline approach to the many facets of networking such as financial considerations, hardware and data transmission considerations, network organization and cooperation, etc. (1-6). Two papers in this category which examine the Iowa Regional Network consider the mutual impact both on the central facility and the remote institutions; the author has used these papers as important background sources for this article (1,2).

To bring our topic clearly into focus, let us couch it in terms of two questions. To what extent can a network meet the needs of a Department of Chemistry of a remote institution? To what extent has the RCC met the needs of the Departments of Chemistry of its member institutions? After a brief résumé of the RCC the paper will examine these needs. They will be divided into two parts, one dealing with the impact of increased hardware flexibility and user-services, and the other dealing with the past, present and future level of computing in the chemistry curriculum.

The Regional Computer Center

Iowa networking began in 1967 when the University of Iowa Computer Center established a telephone line link with two other neighboring institutions. In 1968, with substantial NSF support, the network expanded to eleven institutions: one junior college,

nine four-year colleges, and a government installation. Although
three of these original institutions left the network, the total
today stands at thirteen as listed in Table I. This table summarizes general information about each of the RCC schools. A complete description of the IBM 360/65 configuration, the palm of the
hand located at the University of Iowa at Iowa City, can be found
elsewhere (1).

The RCC from the start has been more than just an organization consisting of faculty and students from the various institutions, computer professionals residing at Iowa City, and computer
hardware/software and communications equipment. It is a dynamic
entity wherein its members assist each other with their primary
effort directed towards instruction. The RCC central staff interfaces with a campus Computer Coordinator at each institution. The
local Coordinator acts as manager of the local terminal, offers
short courses in programming, provides in-house expertise on the
use of the central facility and its diverse compilers, etc. Of no
less significance, he also dons the role of "sparkplug" to excite
local faculty and students to make further use of the whole
computer facility and services.

The University of Iowa Computer Center staff maintains the
hardware and systems on the IBM 360/65. The RCC central staff
links the central facility with the user institutions to achieve
maximum effectiveness in utilization. Beyond the communications
equipment this means regular "update" meetings with frequent on-
site visits by the "circuit riders" to each institution to respond
more directly to particular institutional and faculty needs, and
regular network-wide workshops for the faculty of member institutions. Available to each member institution is access to all the
languages and library program packages available locally at Iowa
City. In addition to FORTRAN, PL/I, COBOL, and ALGOL, these
include the super-batch, in-core compilers WATFIV, WATBOL, PL/C,
ASSEMBLER-G, and SPITBOL which are very fast and inexpensive to
run. Notable is the SPSS package and the Political Science
Department's Social Science Data Archives. Also notable is the
substantial number of computer-based, certified teaching modules
in the disciplines of Business, Chemistry, Economics, Mathematics,
Physics, and Social Science, all because of University of Iowa
participation in the CONDUIT experiment (7).

A Smart Terminal or A Dumb Terminal?

Table I includes a list of the actual terminal hardware as
well as institutional data. The Mason City School District will
be excluded from all further consideration in this study because
it concerns a high school rather than a college system. Most of
the schools have a "dumb" terminal (one which can function only as
a terminal to the central computer facility, such as an IBM 3780),
but several have a "smart" terminal (one with stand-alone academic
as well as remote batch capability, such as an IBM 1130). For

Table I. The Iowa Regional Network Schools

School[a]	Enrollment (Staff)[b]	Terminal	Type
Augustana College Rock Island, IL	2100 (6)	IBM 360/20	Dumb[c]
Central College Pella, IA	1300 (3)	IBM 1130	Smart
Clarke College Dubuque, IA	600 (3)	IBM 1130	Smart
Grinnell College Grinnell, IA	1200 (5)	PDP 11/45	Dumb[d]
Iowa Wesleyan College Mt. Pleasant, IA	600 (2)	DATA 100-70	Dumb
Loras College Dubuque, IA	1500 (4)	IBM 3780	Dumb
Marycrest College Davenport, IA	1000 (2)	DATA 100-70	Dumb
Mason City School Dist. Mason City, IA	e	HP 2000F	Smart
Monmouth College Monmouth, IL	700 (4)	IBM 3780	Dumb
Quincy College Quincy, IL	1300 (3)	DCT 132	Dumb
St. Ambrose College Davenport, IL	1300 (3)	DATA 100-88	Dumb
Univ. of Northern Iowa Cedar Falls, IA	8700 (10)	SPC 16	Dumb[f]
Wartburg College Waverly, IA	1200 (4)	IBM 3780	Dumb

[a] All of the schools listed are private liberal arts colleges except the Mason City public school district and the University of Northern Iowa, a public university.
[b] The number inside the parentheses is the number of staff in the Department of Chemistry.
[c] Although it might serve as a smart terminal, for essentially all academic use Augustana has used its computer as a dumb terminal.
[d] Prior to January 1975, Grinnell had only an IBM 3780. Thus for the period pertinent to this study, Grinnell has only had a dumb terminal.
[e] The Mason City School District will be excluded from any analysis since it is a high school rather than college system.
[f] The SPC 16 can emulate the IBM 360/20; to date it has only been used as a dumb terminal.

reasons given in the footnotes of the table, all but two of the
twelve schools in this study are considered to have dumb terminals.
This classification assumes the reference point of usage to be
academic years 1973-74 and 1974-75.

Three sources of information were used to obtain a measure of
"chemistry computer activities" at the RCC schools. First, RCC
user-statistics from July 1973 through April 1975 were available
from the RCC for each month of the year for each school. These
statistics were categorized according to number and class type of
jobs for each discipline, chemistry in our case. Second, in
addition to the quantitative RCC user-statistics, a survey of the
RCC Coordinators from each school was taken to ascertain chemistry
department usage, user attitudes, coordinator-department cooperation, hardware considerations, etc. Third, a more detailed survey
was taken of each chemistry department to ascertain curriculum
offerings in which the computer was used, the specific type of
use, the level of use, the number of active users on each departmental staff, the sources of chemistry computer programs, etc.
Additional reference to this chemistry questionnaire will be made
in the next section.

One of the options facing a college in Iowa and western
Illinois is whether or not to affiliate with the RCC or to go the
route of a small stand-alone computer. From the perspective of
chemistry users in the Iowa Network, no clear preference arose for
or against a stand-alone computer. To the question "Could a small
stand-alone computer with no Iowa City terminal capabilities more
than adequately serve your present total needs?," the response was
almost evenly divided.

Reasons favorable to a terminal facility varied with the
available peripheral equipment at Iowa City (e.g. disk storage,
calcomp plotter output), the availability of assistance from
computer experts at Iowa City, and the resources at the University
of Iowa available to RCC Chemistry Faculty (e.g. Chemical Titles
files, E.R.I.C. files). Not insignificantly, several chemistry
users opposed a small stand-alone computer because they needed a
sophisticated, large-core computer like the IBM 360/65 for their
research activities.

Reasons favorable to a stand-alone computer obviously rest in
each chemistry user's perception of "needs" to the above question.
For several departmental staffs where only one member used the
computer and only then occasionally, classroom usage was commonly
limited to several canned programs. Not surprisingly this sort of
user sees no broader implications or usage for his students or
himself in the resources available through the RCC. For example,
the use of the computer in a modern chemical literature course has
probably never received serious consideration. Then there is
another situation where a stand-alone facility is satisfactory
under present departmental budget limitations. If most of the
desired classroom/laboratory applications can be accomplished on
the in-house ("free") facility, why use the open-ended ("not free")

terminal and strain the budget? An astute and dedicated Coordinator can readily convince users and administrators that in-house use of elementary computer programs in a number of academic disciplines can be more cost-effective than terminal use.

Clearly there are many factors involved in any institutional decision concerning a stand-alone versus a terminal facility and it is not the purpose of this paper to consider them in detail. But one very interesting trend has clearly emerged from the RCC user-statistics. The relative academic versus administrative use has decreased significantly in over half of the RCC schools. This increased administrative activity is due to the rapid and intentional development in Administrative Data Processing (ADP) systems by the RCC staff. Currently over half the RCC schools use the Registration-Grade Reporting and Payroll systems; member school interest in the use of the Alumni and Business Office systems is growing. It appears at this point that newer and better ADP applications via the terminal to Iowa City will dictate increased rather than decreased support of the terminal for most of the RCC schools.

What about the needs of a Department of Chemistry whose principal concern is limited to the baccalaureate degree? Based on the responses to the chemistry questionnaire, the RCC chemists do not view the provision of large-scale computing capability as primary, except in very few circumstances. Nor do many of them care very much about extra computer and chemistry services available through the Iowa Network arrangement. For many of them a small stand-alone computer will serve just fine in their classroom situation. Yet, in light of the apparent increased total institutional support of a terminal arrangement, the old question of a remote terminal versus a stand-alone computer no longer seems appropriate. Rather the question now seems to be one of whether to select a smart or dumb terminal. A smart terminal configuration would seem to provide the best of both worlds for all disciplines since the desirable administrative and academic applications of a normal remote job entry terminal are still possible while the cost-effectiveness of running many small academic canned programs locally is maintained. Because of particular circumstances and personnel at Central and Clarke Colleges (the only RCC colleges possessing smart terminals used academically), it is inappropriate to use their chemistry usage as indicative of the norm. Yet is it not surprising to find little chemistry terminal usage for either college and also a noticeable increase in terminal ADP type of usage for one of them.

One other very important trend surfaces from both the Coordinator survey and the chemistry questionnaire: there is broad interest in time-sharing among chemistry users. Even those users who could live with a stand-alone, batch-mode computer were interested in time-sharing! As indicated in Table I, Grinnell College is already into time-sharing with the purchase of their PDP 11/45; by fall the facility is expected to be fully operational for

classroom use. The University of Northern Iowa has purchased an
HP 2000F with delivery time this summer. Other individual schools
have made decisions to go into time-sharing and await only the
concerted financial campaign necessary to secure funds for such a
facility. Much of this time-sharing interest is due to the effort
of the RCC staff to involve the network in an effort to purchase a
computer like the HP 2000F, with the cpu located at Iowa City and
the terminals at the participating network institutions. Included
in the plan was a provision for RCC faculty to locate/write software for selected disciplines (one of which was chemistry) to
assure an immediate and fruitful evaluation of time-sharing on a
network-wide basis.

In the hope of proposing a model configuration best satisfying the many and diverse needs of a "typical" smaller Department
of Chemistry, from the experience of the Iowa Network it appears
that a smart terminal arrangement where time-sharing is a part of
the configuration would be the ideal model, one similar to the
facility at Grinnell. However, financial resources do not allow
many (if not most) medium-sized institutions at this point in time
to seriously contemplate such an arrangement.

Program Availability and Level of Use in the Chemistry Curriculum

From the discussion in the previous section, the RCC has had
little or no impact on the member institutions, at least so far as
hardware and "compute power" in the discipline of chemistry are
concerned. Although the use of chemistry services at present is
limited, more and more RCC users are becoming interested in
Chemical Abstracts, Chemical Titles, and E.R.I.C. files. And
there is a definite and broad surge of interest in time-sharing,
much of which is due to RCC staff efforts in convincing users that
RCC time-sharing capability belongs to the "realm of the possible."
Yet these areas have been the least critical to chemistry users
and if we are to fully measure the impact of the RCC on the network Departments of Chemistry we must look elsewhere. It is in
the area of providing programs for use in the chemistry curriculum
that the Iowa Network has made its biggest impact on chemistry
users, and also where there still remains the greatest potential.

It is obvious that there is no way to benefit from computerbased materials in the classroom unless the materials are readily
available. Although the chemistry users in the Iowa Network are
not the worst of the lot classified by Joe Denk (8) as "softwarestarved little people," the difficulties in obtaining good
programs for classroom use are still common to all of us in the
network. In the last two years things have improved considerably.
Typically the user had two sources for obtaining chemistry
computer programs, write his own and/or locate them via the literature. Almost all of the active users continue to do both, but
due to University of Iowa participation in the cooperative national CONDUIT experiment in the transportability of computer-based

undergraduate curriculum materials, the users of the network now also have available CONDUIT as a source of programs.

A quick glance at the chemistry questionnaire and the types of literature sources employed by the RCC chemistry users indicates that the Journal of Chemical Education and CONDUIT are the most common sources of chemistry programs (9). Although readily available on the disk at Iowa City, CONDUIT chemistry programs have suffered in the past from occasional bugs and incomplete or inappropriate documentation. On the other hand, anyone who obtains a program directly from an author noted in the literature does not have the CONDUIT review/certification process to rely upon. Moreover, he must often modify the program to fit his particular computer system. Additional programs have been obtained from the textbooks by Wiberg (10), Dickson (11), DeTar (12), and Isenhour and Jurs (13); from CCUC (14); from the Quantum Chemistry Program Exchange at Indiana University; and from Oak Ridge and Argonne National Laboratories through the University of Iowa Department of Chemistry (15,16).

According to the twelve Departments of Chemistry of this study, the use of the computer occurred most often in the physical chemistry curriculum (eight users indicated regular use on the average of at least once a month). Computer usage in analytical chemistry was next most frequent (only three users indicated no use at all), followed closely by freshman chemistry and then organic chemistry (six users indicated no use at all). Other curriculum offerings in which the computer was used ranged from courses in the Identification of Organic Compounds and Biochemistry to those in Research and Independent Study and one in Computer Applications in Chemistry. Types of programs vary from least squares analysis and other data reduction routines to plotting and other simulation programs. A few of the specific applications include kinetics studies, quantum mechanical calculations, numerical integrations, atomic and molecular orbital calculations, IR and NMR routines, and several titration applications (from simulation to endpoint determination).

In addition to the increased base of programs available for future use in the chemistry curriculum, another important benefit from network affiliation is the holding of network-wide workshops and meetings. The intent of such workshops has been to foster interaction between members of remote institutions and stimulate new ideas and applications in the classroom. As a point of reference, the last chemistry workshop was help in September 1973. About thirty participants (chemistry faculty and students and some Coordinators) attended from eleven of the RCC institutions and also from three other Iowa colleges and universities. Over two-thirds of those schools which were both at the workshop and are also a part of this study claimed that the workshop increased both faculty interest in new computer programs and usage of the computer. The user-statistics indicated a surge of chemistry use in the several months immediately following the workshop.

And of the twelve schools in this study, all but one wanted at least occasional chemistry workshops, with nearly one-half wanting annual workshops!

From earlier discussion it should be clear that the RCC has functioned more as a "user-services" network than a "transmission" or "facilitating" network (4). The fingers and the palm of the RCC hand seem more and more to realize their mutual interdependency. It is the author's feeling that there is a growing change in attitude among chemistry users. In their eyes the RCC has always existed to provide a facility for accomplishing specific computational tasks. But with workshops by and for chemical educators in which pedagogical use of the computer becomes an important ingredient, the network becomes more than just an agent to parcel out bigger and better canned programs. Such workshops can elicit interest from the uninitiated as well as serve as a sort of fueling station for the more knowledgeable users. In addition to providing new and useful knowledge about particular routines, the interpersonal contacts of the workshop more often than not serve to kindle further activity in the participants. As a case in point, the frequent workshops held by the North Carolina Educational Computing Service have played a key role in the high rate of growth in the use of the computer in the North Carolina Network (17). To be sure there are still chemistry users who are not interested in network sharing in the Iowa Network and who would rather "do it alone." Another type of response to our network, the "do it for me" attitude, is also still too common. Yet the author detects a trend away from these responses and instead a "do it with me" response prevailing in more and more chemistry users. The inclusion of more frequent faculty training workshops as an integral part of the overall RCC implementation strategy at present appears to be one of the most effective ways to capitalize on this favorable response. In turn, the increased level of usage would benefit both fingers and palm significantly.

Conclusions

The computer is fast becoming a tool vital to the whole of modern-day chemistry. It is causing a revolution in the frontiers of chemical education! The activity of the ACS Divisions of Computers in Chemistry and Chemical Education (particularly its Committee on the Role of Computing in Chemical Education) at recent Regional ACS Meetings and National ACS Meetings (including this one) is not going unnoticed by the chemists at the RCC member institutions. Although in the recent past the chemistry users have not taken advantage of computing capability and chemistry services afforded through the RCC, they are beginning to recognize them as probable future needs of a modern Department of Chemistry. The present financial situation at most of the RCC schools precludes immediate and extensive use of these features, but the chemistry users do seem to recognize the potential that exists.

Increased availability of computer programs for use in the chemistry curriculum and network-wide chemistry workshops have had a strong impact on the chemistry users of the RCC institutions. Regular chemistry workshops for and by chemical educators are clearly recognized as valuable and highly desired by almost all the member institutions' Departments of Chemistry. The RCC has generated a wholesome demand for network cooperation and with extra effort towards filling this demand, such as more frequent workshops, it is in a position of seeing a significant increase in the level of computing by the chemists of the Iowa Network.

Finally, there is a growing nationwide interest in the use of computer-assisted test construction (CATC), with the discipline of chemistry taking an active leadership position (18). Moreover, the CONDUIT Chemistry Advisory Committee is examining the possibility of providing a transportable CATC item pool and question retrieval program (19). Because of these circumstances and the close-to-home fact that the Department of Chemistry at the University of Iowa has just completed a CATC program in freshman chemistry (20), it is but a matter of time before Iowa chemistry users will have to face the decision of CATC on a network-level cooperative basis, or on an individual institutional basis, or not at all! Because of the considerable effort involved to construct and maintain a large data pool of chemistry questions acceptable to even a few users and because of the size of the computer necessary for even a modest CATC system, a cooperative CATC venture would be a natural next step for a network like the RCC. Although CATC is still off on the horizon, so far as most RCC chemistry users are concerned, a network-wide effort in this area of instructional computing could prove very beneficial to each of the RCC Departments of Chemistry. It should hold a key position in future institutional and network deliberations.

Acknowledgements

It is a pleasure to acknowledge Chuck Shomper, Director of the Regional Computer Center, and Pete Trotter, Manager of Network Academic Services of the RCC. Without their help this study would have been impossible. Willing assistance from the chemistry users and Coordinators of the RCC member institutions is also gratefully acknowledged.

Literature Cited

1. Weeg, G.P., and Shomper, C.R., EDUCOM Bulletin (Spring 1974), 9, 14.
2. Weingarten, F.W., Nielsen, N.R., Whiteley, J.R., and Weeg, G.P., "A Study of Regional Computer Networks," University of Iowa, Iowa City, IA, 1973.
3. See, for example, "Networks and Disciplines" (Proceedings of the EDUCOM Fall 1972 Conference), EDUCOM, Princeton, NJ, 1972.

4. Greenberger, M., Aronofsky, J., McKenney, J.L., and Massy, W.F., Science (1973), 182, 29.
5. Luehrmann, A.W., and Nevison, J.M., Science (1974), 184, 957.
6. Chambers, J.A., and Poore, R.V., Communications of the ACM (1975), 18, 193.
7. CONDUIT stands for Computers at Oregon State University, North Carolina Educational Computing Service, Dartmouth College, and the Universities of Iowa and Texas (Austin). Sponsored by the National Science Foundation, this consortium of regional computer networks was organized in January 1972 to study and evaluate the transportability and dissemination of computer-based curriculum materials for use on the undergraduate level of instruction.
8. Denk, J.R., Proceedings of the Conference on Computers in the Undergraduate Curricula (1972), 3, 547.
9. The Journal of Chemical Education is a very common source of computer programs for all sorts of applications in the chemistry curriculum. See, for example, the "Selected Bibliography of Computer Programs in Chemical Education" submitted by the author for publication in J. Chem. Educ. This bibliography lists 168 digital computer programs noted in J. Chem. Educ. for the eight years 1967-74; computer language and machine and a brief statement about the program usage are given.
10. Wiberg, K.B., "Computer Programming for Chemists," W. A. Benjamin, Inc., New York, 1965.
11. Dickson, T.R., "The Computer and Chemistry," W. H. Freeman and Co., San Francisco, 1968.
12. DeTar, D.F., "Computer Programs for Chemistry," W. A. Benjamin, Inc., New York, 1968(I), 1970(II).
13. Isenhour, T.L., and Jurs, P.C., "Introduction to Computer Programming for Chemists," Allyn and Bacon, Inc., Boston, 1972.
14. CCUC stands for Conference on Computers in the Undergraduate Curricula. Below are dates and places of the annual conference.

    CCUC/1, Iowa City, IA   June 1970
    CCUC/2, Hanover, NH   June 1971
    CCUC/3, Atlanta, GA   June 1972
    CCUC/4, Claremont, CA   June 1973
    CCUC/5, Pullman, WA   June 1974
    CCUC/6, Fort Worth, TX   June 1975

    Copies of the Proceedings for any of these conferences can be purchased from Ted Sjoerdsma, Lindquist Center, University of Iowa, Iowa City, IA, 52242.
15. Johnson, C.K., "ORTEP (Oak Ridge Thermal Ellipsoid Plot Program)," Oak Ridge National Laboratory Report 3794, 1965.
16. Wahl, A.C., Bertoncini, P.J., Kaiser, K., and Land, R.H., "BISON, A Fortran Computer System for the Calculation of Analytic SCF Wavefunctions, Properties, and Charge Densities for Diatomic Molecules," Argonne National Laboratory Report 7271,

1968.
17. Joe Denk, private communication.
18. See, for example, "The Computer Assisted Test Construction Conference" held at San Diego, October 1974.
19. Private communication.
20. Kenneth Sando, private communication.

11

A Case History in Computer Resource Sharing: *ab initio* Calculations *via* a Remote Control*

D. G. HOPPER,† P. J. FORTUNE, and A. C. WAHL
Chemistry Division, Argonne National Laboratory, Argonne, Ill. 60439

T. O. TIERNAN††
Chemistry Research Laboratory, Aerospace Research Laboratories,
Air Force Systems Command, Wright-Patterson Air Force Base, Ohio 45433

In this paper we discuss the experience that we have had in performing large scale molecular structure calculations remotely using the CDC6600 computers at Wright-Patterson Air Force Base. Certain elements of this effort are general and can be expected to be encountered by other researchers. Before proceeding with a discussion of this computational research project we will give a brief history of the development of the Wright-Patterson AFB Aeronautical Systems Division (ASD) computer center, which played a historic and central role in the development of modern computational chemistry.

Computational chemistry has become a pervasive tool contributing to the solution of problems in the biological, inorganic, organic, and physical subfields of chemistry. And it will undoubtedly become an even more important tool in years to come. Such growth is due to the fact that in many areas of computational chemistry, the methods used yield results which meet experimental accuracy, allowing quantitative prediction and interpretation to be made (1-3). Methods of performing e.g. accurate quantum mechanical calculations are now available in computer codes which can be obtained from code exchanges (3) and from individual research groups. It is proper to refer to these codes and the hardware associated with their application as instruments for chemical research, in the same sense as an

---
*Work performed under the auspices of the USERDA and Air Force Contract No. F33615-72-M-5015.
†ARL-NRC Research Associate 1972-1974.
††Present address: Department of Chemistry, Wright State University, Dayton, Ohio 45431.

153

equally sophisticated experimental "apparatus" (4).
Remote usage of these large computational systems is
becoming increasingly more common. The reasons for
this increased remote usage are: (a) true portability
is difficult to achieve in a large software system, (b)
the maintenance, reliability, and continuing development of these codes requires more effort and expertise
than can be marshalled by individual chemical researchers and (c) remote use is possible, convenient, and
economically competitive with on-site visits. It is
reasonable to expect that an even more substantial portion of chemical calculations will be performed remotely from interactive (time-sharing) and batch terminals
(5), particularly as computer networks (6, 7, 8, 9)
become more widely accessible.

In this report we review our specific experience
with remote interactive terminal access. We discuss
the feasibility of performing moderately extensive ab
initio production calculations with no more than a
teletype at the remote site. The entire procedure is
straight-forward and, although tedious, can be greatly
facilitated by appropriate code modifications and code
developments. The following discussion is broken down
into the six sections: nature of the calculations
being performed, equipment available, communication
links, installation and maintenance of codes, input and
output of jobs, and a summary and indication of future
plans.

For the purpose of gaining perspective on this effort we begin by reviewing the role played by the ASD
Computer Center at Wright-Patterson AFB in computer
resource sharing in quantum chemistry computations.

History of Computer Resource Sharing at Wright-Patterson Air Force Base

The historical evolution of our remote usage has
paralleled the development of the computer center at
Wright-Patterson Air Force Base. It was the sharing of
computer resources by Wright-Patterson in the late
1950's that enabled one of the first integral programs
to be developed and run. This evolution in the use of
the Wright-Patterson computational facility by off-site
personnel, accomplished entirely by site visits in
about 1958, to the present stage of extensive remote
operations provides, therefore, a vivid example of computer resource availability and sharing. This example
is illustrative of a research area--computational chemistry--which will be markedly advanced by the development of computer networks.

Background on the ASD Computer Center. There has been a computer organization as a part of the Wright-Patterson Area B laboratory complex since 1949 (10). At that time the Computational Branch of the Research Division, Office of Air Research, was established. This group began with an early vintage MIT analog computer which was soon replaced with the first commercially available analog computer, the REAC Series 100. The digital effort began in 1951 with the delivery of an IBM Card Programmed Calculator. To put these events in perspective, the digital computer was invented by Harold Aiken only in 1944; the analog computer, by Vannevar Bush in 1930 (11).

The first real stored program digital computer was the OARAC (Office of Air Research Automatic Computer) installed in 1953. This computer was a "one-of-a-kind" built specifically by General Electric for the Aeronautical Systems Division. It was very slow, had limited I/O capability, and was unreliable. In 1956 it was replaced by a scientifically-oriented Univac 1103. In 1957 the 1103 was upgraded to an 1103A and assembly language programming became available. It is this computer that played an important roll, as described in the next section, in the development of ab initio quantum chemistry.

The first extensive use of a source language didn't occur until 1961 when the 1103A was replaced by an IBM 7090 which allowed programming in FORTRAN II. Open shop was formed at about this time. The IBM 7090 was the first real batch computer at ASD; it used two IBM 1401 computers for converting cards to tape input for the 7090 and 7090 tape output to punch and printed form. In 1963 the 7090 was upgraded to a 7094 and then replaced in late 1964 by an IBM 7044/7094 Mod II Direct 7044 to monitor I/O, job flow, and disk storage allocation for the 7094 program execution. This system had one IBM 1440 remote batch terminal in the Aero Propulsion Laboratory. In 1966 a second, somewhat slower IBM 7040/7090 Direct Coupled System was added; it had a remote batch terminal in the Flight Dynamics Laboratory. Both Direct Coupled systems were replaced by a CDC6600 with 31 teletypes and 9 remote batch terminals in January 1971. In December 1973 a CDC CYBER73 was installed with the batch terminals reallocated among the two systems.

The teletype terminals (110B)* were converted to operate through a dial-up system in March 1973 and some 300B lines were simultaneously added. There are now 32 110/300B lines into each system, the CDC6600 and the CYBER73. One line operating at 2 K/B was added in

*Notation -- B = Baud = bit-per-second. 1K/B = 1000B.

April 1974 and a second operating at 4.8 K/B, in February 1975. The ASD Computer Center became a node on the ARPA network (ARPANET) in October 1973 with the installation of a terminal interface processor (TIP).

Future developments planned for the ASD and other Air Force System Command computer centers will be discussed below. For now we turn to a discussion of the role played by the 1103 and 1103A in the evolution of quantum chemistry.

Development of Master Integrals and SCF Codes. During the period 1955-1962 the master SCF molecule program was developed for linear systems by the Laboratory of Molecular Structure and Spectra of the University of Chicago as a part of its overall effort in molecular computations (12,13). The molecular integrals package was developed and run on the Univac 1103 and 1103A at Wright-Patterson Air Force Base under a contract with the University. The SCF package was combined with the STO integral package to form the first master integrals-SCF code. It was in this period of calculations on the 1103 and 1103A that many of the first molecular Hartree-Fock calculations with good basis sets were carried out by McLean, Weiss, and Yoshimine (14), Kolos and Roothaan (15,16), Ransil (12,13), and Richardson (17), etc. Indeed one finds that the April 1960 issue of Reviews of Modern Physics -- a collection of papers given at the Conference on Molecular Quantum Mechanics held at Boulder, Colorado June 21-27, 1959 -- contains no fewer than twelve papers that acknowledge the use of the 1103 at the then Wright Air Development Center at Wright-Patterson AFB (12-23). These papers include contributions by Kolos, Roothaan, and Sach (18) on the ground state of H_3, Roothaan (19) in his classic work on the theory of open shells of electronic systems, Kolos and Roothaan (20) on correlated orbitals for He, Fraga and Mulliken (21) on the role of Coulomb energy in valence bond theory, Fröman (22) on relativistic corrections, and Löwdin (23) on expansion theorems for the total wavefunction and extended Hartree-Fock schemes. Many other individuals participated in this calculative effort at Wright-Patterson--Bagus, Clementi, Ehrenson, Huo, Lykos, Malli, Phillipson, and Wahl to mention a few. S. Huzinaga writes in a letter to Dr. R. Euwema of the theoretical solid-state group at Wright-Patterson dated February 7, 1975: "Wright-Patterson is one of my fondest memories during my stay in U.S.A. some 15 years ago. I had access to a big computer (UNIVAC 1103?) for the first time in my life."

The 1103 and 1103A provided a computational test ground for many ideas in theoretical chemistry and a place where the seeds of future developments in the 1960's were sowed. It was here, for example, where newer ideas on how to code diatomic integrals were evaluated and the one-electron integrals coded and tested (24).

The chores of coding and execution of computations on the 1103A were markedly different from the present-day high-level language and batch job input operations, as one of us (A.C.W.) vividly recalls. Coding was performed directly in octal machine language, to optimize the use of computer memory. Accurate floating-point operations required a user-supplied subroutine. Job submission was an extraordinary exercise in preparation and patience. The computer resided on a lovely, blue-lit dias and the user waited in a "ready room" until his name was called. He would then rush in with his paper tapes for a 2-3 minute shot at getting his job on the machine. If he failed to get it going in that length of time he lost his turn and had to go to the end of the queue. Another turn would come in two hours to two days, depending upon demand.

The use of the 1103 and 1103A by the Chicago group was a milestone in the early sharing of computer resources. In this case computational chemists obtained access to a sufficiently powerful computer for their purposes - a computer unavailable to them in Chicago - by making site visits to Wright-Patterson.

Ab Initio Quantum Calculations from 1961-1975 at Wright-Patterson AFB. During the period 1961-1972 the ab initio work in quantum chemistry at Wright-Patterson AFB was much more limited than the 1956-1961 period. The reason was the availability of equivalent or better computers at or very near the institutions in Chicago and elsewhere at which the quantum chemists resided. However, during much of this period and continuing on until 1975 there has been a strong theoretical solid state group at the Aerospace Research Laboratories. This group has been developing methods and very highly sophisticated codes for performing rigorous ab initio Hartree-Fock calculations over the years on the IBM 7094 and, currently, the CDC6600.

The period 1972-1975 saw an upsurge in ab initio quantum chemical calculations. A few group leaders in the Chemistry Research Laboratory of the Aerospace Research Laboratories brought in theoretical personnel on a temporary basis to execute computational research projects related to on-going, in-house experimental programs. The Aerospace Research Laboratories, through a program initiated by one of the present authors (T.O. T.), also contracted out-of-house theoretical calculations relevant to Air Force interests. One such contract involved the computational chemistry groups at Argonne National Laboratory and the National Bureau of Standards who undertook a concerted literature evaluation and initiated a state-of-the-art computational project to obtain information on the vertical excitation spectra and potential energy hypersurfaces of the electronic states of the fifteen molecules and ions H_2O^m, N_2O^m, CO_2^m, NO_2^m, O_3^m, m = +1, 0, -1 (24-31). The work reported in this paper is a development of that computational project. It is in the context of this program that we will discuss remote usage.

Nature of Project and Calculations Being Performed.

For about three years our research groups, in concert with those of M. Krauss at the National Bureau of Standards (NBS) and J. Simons at the University of Utah (Utah) have cooperated in computational chemistry projects. In these joint investigations the group at the Aerospace Research Laboratories (ARL) collaborated in the development of objectives, monitored progress, and provided computer resources. The group at Argonne National Laboratory (ANL) constructed CDC6600 versions of its BISON, BISONMC, DASCI, and POLYINT codes* and made them operative at Wright-Patterson AFB (24-28). The group at the University of Utah implemented codes for its equations-of-motion method for computing ionization potentials and electron affinities from SCF wavefunctions (32). Our remote interactive terminal set-up is depicted schematically in Figure 1. These research groups then proceeded to pursue program objectives utilizing these common resources.

The ARL-ANL-NBS project is designed to catalog what is presently known from both theory and experiment about the dominant atmospheric molecules H_2O, NO_2, CO_2, O_3, N_2O and their positive and negative ions, to critically review this information, and to supplement that information in a systematic computational fashion with theoretical calculations. This new theoretical knowledge can then be used to advance the experimental

---
*This set of codes provides for the ab initio computation of polyatomic wavefunctions, properties, and potential surfaces by the optimized valence configurations (OVC) multiconfiguration self-consistent-field (MCSCF) configuration-interaction (CI) technique.

analysis by participating in a feedback process which historically has proven to be important in understanding the phenomena involved. A critical review of the literature has been completed in which extensive use is made of state adiabatic correlation diagrams to summarize potential energy surface characteristics (33). The vertical spectra of the above fifteen molecules and molecular ions have been computed at the SCF level and, for some, at the OVC-MCSCF-CI* level (34). Potential surfaces for various states of H_2O^+, H_2O, N_2O^+, N_2O, N_2O^- and NO_2 have also been examined (34). Further studies with the OVC-MCSCF-CI technique are currently underway.

Terminal Equipment

In the early stages of this project all activity proceeded by site visits to Wright-Patterson Air Force Base (WPAFB). After about one year, dial-up interactive terminal connections became available at WPAFB and we were able to perform much of the maintenance and production work remotely from 110B and, somewhat later, from 300B interactive terminals. Most of our effort was with simple 110B teletypes located at ANL, NBS, Utah, and WPAFB. A 300B CRT terminal was available for code maintenance on-site at Wright-Patterson on the same dial-up basis as the 110B and 300B off-site terminals.

The requirements imposed upon the choice of a remote terminal and upon the tactics of operation in an effort such as ours are essentially those of any large scientific computational project. One needs the capacity to formulate and transmit numerous production jobs to the host computer batch input queue and to receive and examine output. It is then necessary to be able to manipulate files on the host computer via some sort of an interactive option (e.g. CDC INTERCOM, IBM TSO). However, there are some characteristics of ab initio quantum chemical calculations which require special attention. While it might be possible, for instance, for us to establish production load modules periodically, it is necessary to maintain and continually upgrade each of five, 4000-7000 card, source codes resident on the host computer. A CRT terminal component and a means of obtaining updated line-numbered, 80-80 source listings worked well for us. This procedure required on-site personnel. Some modifications were accomplished

---

*See footnote on previous page.

completely remotely with a teletype from either Argonne
or Utah. Furthermore, the input decks for production
runs are quite large (200-400 cards) and often change
very substantially from one run to the next. For this
reason it is not expedient to construct these input
decks solely via the keyboard. Instead, we found it
useful to maintain a library of model input decks on
the host computer mass storage media or on remote site
media such as paper tape or magnetic tape cassettes.
A model input deck can consist of the "punched" output
from a previous run and may be stored with the job control cards attendent to its execution in place. Input
decks on the above-mentioned remote site media can be
constructed locally from punched cards if a batch computer facility is available. The text editor of the
interactive software system of the host computer can
then be used to construct the desired input deck from
these sources.

Another characteristic of the type of calculations
with which we are concerned is the large size of the
output files. For such files as these -- files often
in excess of 4000 lines -- complete printing at a remote interactive terminal operating at 110-1200B is not
practical. Even a remote batch terminal operating at
2000-4800B would require prohibitive transmission times.
To remedy this problem we have modified our codes to
produce printed output files in a summary, as well as
the detailed standard, form. For production calculations with our BISONMC code some 100 70-column lines
suffice to contain the important information. A teletype has served, in our project, to enable the production of such printed summaries at each remote site.
The standard output files were printed at the host computer and mailed to the respective user as needed.

Table I is a comparative display of the characteristics of the interactive terminals we have used
remotely -- units 1 and 2 -- and would like to -- unit
3. Some desirable batch terminal characteristics are
included in Table I for completeness; discussion of
these is deferred to a later section. For our type of
application, options such as disk packs, plotters, and
tape drives, while costly to acquire and maintain, provide no service not already available via the host computer. For this reason they have been omitted from
Table I. Of course, if some of these components are
already available, they can be used to advantage. Also,
a remote site text editing capability can be helpful to
our type of application. Examples of the latter include a dual tape cassette unit and the interactive
software of a computer local to the remote site.

Unit 1, the teletype or equivalent device, is prominently numbered in Table I to stress the facts (a) that it is in our experience capable of enabling, at minimal cost, a modest rate of production computation and (b) that it is already widely available. For individuals evaluating equipment for a remote terminal (interactive or batch) we note that a very detailed compendium of information about terminals, modems, data communications, etc. is available (35) as well as other background information (36).

Communication Links.

The most common communication link between remote terminals and computers is a phone line. While the simplest and an often reliable way to use a phone line is to direct-dial over the public network, this is not always possible. Thus, it is often necessary to go through extra layers of effort on each call -- i.e. institutional exchanges -- and to put up with the noisy lines of private phone networks. The higher transmission failure rates on noisy lines can limit an interactive terminal/host computer combination capable of 1200B or 300B to 110B. Regular voice-grade public telephone lines are capable of transmission rates up to 4800B (35). Put another way, the undetected error rate on public lines is one in 10^5-10^7 bits at most, given the current technology for signal transmission over an audio wire. Leased lines are much more expensive but can support up to 9600B (35). Interactive terminals and host computers are commonly capable of supporting 110, 300, and 1200B operation. Remote batch terminals and host computers are commonly capable of supporting 2000 and 4800B operation. However, it is clear that one should not purchase a terminal and modem capable of more than 110 or 300B if it is not possible to establish that the remote-to-host phone connection can support the higher transmission rate.

When our project began we were limited to 110B transmission by our interactive terminal -- a Model 33 Teletype -- and we employed direct-dialing of the Wright-Patterson AFB computer from Argonne. Later we upgraded our interactive terminal capacity to 300B but were still limited to 110B by a requirement that we employ FTS lines, which we found to be rather noisy. This problem was compounded by the fact that FTS calls to the computer at Wright-Patterson had to go through the base switchboard and were, therefore, limited to about five minutes during business hours. Future solutions to the communication link problem (cost and

Table I. Remote Terminal Choices for Computational Chemistry.

Unit	Baud Rate Cap.	Equipment[a] Type	Usage In	Usage Out	Stor.	Cost
1	110	teletype				~$ 900
		keyboard	x	-	-	
		70 col. printer (impact)	-	x	x	
		paper tape	x	x	x	
2	110	electronic				~$ 3600
	300	keyboard	x	-	-	
	1200	80 col. printer (thermal)	-	x	x	
		dual tape cassettes	x	x	x	
3	110	electronic				~$ 9000
	300	keyboard	x	-	-	
	1200	132 col. printer	-	x	x	
	2400	dual tape cassettes	x	x	x	
	4800	card reader	x	-	-	
		card punch	-	x	x	
		CRT	x	x	-	
4	2000	batch terminal				~$30000
	4800	keyboard	x	-	-	
	9600	132 char. printer	-	x	x	
		card reacher	x	-	-	
		card punch	-	x	x	

[a]Obtention of an appropriate modem is assumed (35).

transmission rate) and alternatives to phone lines will be discussed below. From our experience on communication links the phone connection is the major difficulty of remote data processing.

Installation and Maintenance[*] of Codes

In our work the computer programs (codes) have been installed only by site visits. Maintenance[*] of

---
[*]Maintenance of codes is, in the present context, taken to include not only the maintenance of operative load modules and source decks on the host computer public disk and tape libraries, but also the updating of codes to accomodate changes in the operating system, to improve performance, and to incorporate new features.

the four ANL codes (24-27) was also performed on-site for the most part, mainly by the use of a CRT interactive terminal. The latter was greatly facilitated by the generation of an updated source listing, line numbered in the same style as in the edit file, after each code modification session at the terminal.

However, a significant portion of the code maintenance was performed via teletype from either Argonne or Utah. It was very helpful, therefore, to keep a current and detailed set of manuals for the host computer at each remote site (Argonne and Utah) from which code maintenance activity was undertaken.

Input and Output

Job Input. The traditional way of job input is by physical card decks via a card reader. A cardreader can be obtained as an add-on option to many interactive terminals on the market and is a standard component of batch terminals. However, as discussed earlier a card reader is not indispensible to an interactive terminal operation. The power of the host computer interactive software (e.g. INTERCOM on CDC and TSO on IBM machines) can be brought to bear on job assembly; and dual tape cassettes can make it possible to perform many of the same job construction tasks offline. It is helpful to modify codes to accept formatfree input as ours have been (24-27). In addition, small programs for, (a) data set construction-by-interactive-terminal-interview, and (b) data set verification, are quite useful. We have written such small, rapid turn-around programs for use in conjunction with our BISON and BISONMC codes (24,25). Such user software modification, development, and utilization measures make it possible to operate efficiently without physical card decks.

In the presently reported work jobs were established in the edit file under the EDITOR mode of CDC INTERCOM and then saved and batched into the input queue. The edit file was established by (a) key-in, (b) read-in of a paper tape, or (c) load-in of a file stored on a host computer disk, followed by appropriate modifications. At Argonne, a utility program was written to run on the Chemistry Division SIGMA 5 computer to convert a punched-card input deck for the CDC6600 into a paper tape. This paper tape was then read in over the teletype. Back-up of input jobs was on paper tape or host disk file.

Output -- General Comments. Output files are generated for several reasons: the perusal of results from a run, the construction of input decks, and the archiving of results. As with the input of jobs, output has traditionally been hard copy -- print-outs and punched card decks. Again, both of these are usually available from a batch terminal. Alternatively, one can take advantage of the host computer interactive software and coding innovations to handle the necessary output manipulations from an interactive terminal. This is especially advantageous for long print-files, for large numbers of print files, and for punch files. Any portion of a file can be listed at will, so that there is less need to print full output files. An output mail-back procedure must still be maintained, however, for those cases where a lengthy print-out (on paper or microfiche) or real punched output is necessary. The perusal function can be satisfactorily performed with a CRT output display or a low-speed printer. The archival function requires some type of storage device on the host computer or at the remote site. File storage and manipulation are discussed under other subheadings.

In our case all remote output has been by teletype or teletype-like devices. It was necessary to store the punch file as a catalogued disk or magnetic tape file during execution of a remote-entry job on the WPAFB CDC6600; otherwise it was lost to the remote user. Also, output print and punch files had to be backed up on host computer tape before run termination if we were to avoid suffering from the occasional loss of unarchived and of temporary disk files. We protected ourselves from losing a run completely, due to, (a) abnormal program termination or (b) an operator drop (e.g. "SORRY -- MUST DROP TO RUN CLASSIFIED") by making the punched as well as the printed output summary file dynamic in that it was updated after each iteration to the latest orbitals, energies, and configuration coefficients.

Output -- 70 Character/Line Option. Because narrow carriage printers and CRT screens are so common for interactive terminals, we have modified our codes POLYINT, BISONMC, and DASCI to provide an option by which the user may specify that all printed output files be in a 70 character/line format.[*] These

---
[*] We chose 70 characters/line because we found it to be the maximum suitable for a model 35 tty.

11. HOPPER ET AL. *Calculations via Remote Terminal* 165

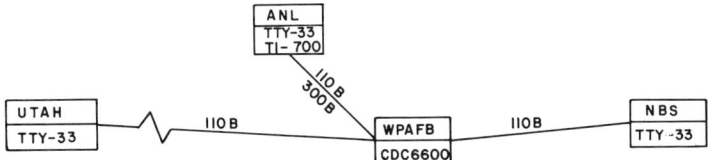

Figure 1. *Interactive terminal connections to Wright-Patterson AFB from Argonne National Laboratory, The National Bureau of Standards, and the University of Utah*

```
PAINT -- ANL MODIFICATION OF THE PA300 POLYATOM PROGRAM.
         INTEGRALS ARE GENERATED IN CANONICAL LISTS FOR
         INTERFACE WITH THE ANL CODES OVC, KRAG, AND DASCI.
BY M.BASCH, BELL LABORATORIES
PA300/CDC6600 BY C.J.HORNBACK, NYU.
ANL CANONICAL VERSION BY A.HINDS, ANL.
ANL VERSION IMPLEMENTED ON CDC6600 BY R.P.HOSTENY, ANL.
ANL VERSION/CDC6600 MODIFICATIONS BY D.G.HOPPER, ANL.
THIS IS THE WPAFB CDC6600 CANONICAL VERSION OF APRIL,1975
FOR INFORMATION OR THE LATEST WRITE-UP CONTACT
D.G.HOPPER, ANL, XARL/LJ, WPAFB,OH.45433, PH513-255-4869.
***************************************************

DATE -- 05/07/75      TIME -- 17.31.12.     DATA SET   1.

RUN TITLE
NNO 843 P111

PROGRAM CONTROL OPTIONS
 0 1 0

NUCLEAR CENTERS
CENTER LABEL -------------COORDINANTS------------- CHARGE
    1   N1    0.00000000  -2.13198906   0.00000000   7.
    2   N2    0.00000000   0.00000000   0.00000000   7.
    3    O   -.00000000   2.23781372   0.00000000   8.

GAUSSIAN FUNCTION SPECIFICATIONS
NUMBER OF PRIMITIVE GAUSSIANS =  72
NUMBER OF BASIS     FUNCTIONS =  39
BASIS SET (SYMMETRY) BLOCKING =  21   18

GAUS FUNC COMP CENT TYPE      EXPONENT     COEFFICIENT
   1    1    1   N1    S     5909.4400000   .0020040
   2    1    2   N1    S      887.4510000   .0153100
   3    1    3   N1    S      204.7490000   .0742930
   4    1    4   N1    S       59.8376000   .2533640
   5    1    5   N1    S       19.9981000   .6005760
   6    1    6   N1    S        2.6860000   .2451110
   7    2    1   N1    S        7.1927000  1.0000000
   8    3    1   N1    S         .7000000  1.0000000
   9    4    1   N1    S         .2133000  1.0000000
  10    5    1   N1    Y       26.7860000   .0382440
  11    5    2   N1    Y        5.3564000   .2438460
  12    5    3   N1    Y        1.7074000   .8171930
  13    6    1   N1    Y         .5314000  1.0000000
  14    7    1   N1    Y         .1654000  1.0000000
  15    8    1   N2    S     5909.4400000   .0020040 CPD FROM FUNC  1
  16    8    2   N2    S      887.4510000   .0153100 CPD FROM FUNC  1
  17    8    3   N2    S      204.7490000   .0742930 CPD FROM FUNC  1
  18    8    4   N2    S       59.8376000   .2533640 CPD FROM FUNC  1
  19    8    5   N2    S       19.9981000   .6005760 CPD FROM FUNC  1
  20    8    6   N2    S        2.6860000   .2451110 CPD FROM FUNC  1
  21    9    1   N2    S        7.1927000  1.0000000 CPD FROM FUNC  2
  22   10    1   N2    S         .7000000  1.0000000 CPD FROM FUNC  3
  23   11    1   N2    S         .2133000  1.0000000 CPD FROM FUNC  4
  24   12    1   N2    Y       26.7860000   .0382440 CPD FROM FUNC  5
  25   12    2   N2    Y        5.3564000   .2438460 CPD FROM FUNC  5
```

Figure 2(a). *Examples of 70-column output formatting for teletype and teletype emulating terminal devices. Only partial listings are shown. (a) PAINT: an unoverlayed CDC6600 version of the ANL code POLYINT.*

```
OVC-1 --- A COMPUTING SYSTEM FOR MCSCF CALCULATION
          OF OPTIMIZED VALENCE CONFIGURATION WAVEFUNCTIONS
          FOR POLYATOMIC MOLECULES
BY G.DAS A.HINDS AND A.C.WAHL, CHEM. DIV. ARGONNE NATIONAL LABORATORY
IMPLEMENTED ON THE CDC6600 BY R.P.HOSTENY, ARGONNE NATIONAL LABORATORY
MODIFIED BY D.G.HOPPER AND P.J.FORTUNE,   ARGONNE NATIONAL LABORATORY
THIS IS THE CANONICAL VERSION OF APRIL,1975.   FOR THE LATEST WRITE-UP,
CONTACT D.G.HOPPER, %ARL/LJ, WPAFB,OH45433, PH513-255-4859.
****************************************************************************

DATE -- 05/08/75     TIME -- 19.50.24.      DATA SET  1.

RUN TITLE
          NNO 1SI+ B43 P111

PROGRAM CONTROL FLAGS
IOPEN=   0   IVECT=   0    IPRINT=   0    KFLAG=    4    JFLAG=   4
IPRCT=   0   ISCF=    0    ICH=      0    IPUNCH=  -7    IDEBUG=  0
INFLG=   0   ISAVE=   1    IOVC=     1    NORBPB=   6

DOCUMENTATION OF INTEGRAL GENERATION FROM BISON STO BASIS FUNCTIONS
ZA    1.000                 ZB   60.823        XNUCL      60.8223803200 A.U.
INTERNUCLEAR DISTANCE            1.0000 BOHRS  NO.GAUSS POINTS ON XI    30
INTEGRATION TRUNC. AT            1.0000 BOHRS  NO.GAUSS POINTS ON ETA   30
NDUM                                  0        MDUMF                     0
NO.NEUMANN ITERATIONS                 0        NO.SIMPSONS POINTS        0
MASS OF CENTER A               778.0000        MASS OF CENTER B.   111.0000
VECTOR COEF.DIAG.THS.          1.000E+00       BIAS FOR ONE-ELEC.GRID    0
COULOMB-HYBRID CONV.THS.       1.000E+00       PRIOR ITERATIONS          0

INTEGRALS OBTAINED FROM SAVE TAPE = UNIT 43
JSUPMX   INTEGRALS COPIED FROM UNIT 43 TO ESTABLISH DISK DATA SET 40
KSUPMXS  INTEGRALS COPIED FROM UNIT 43 TO ESTABLISH DISK DATA SET 41
UTS      INTEGRALS COPIED FROM UNIT 43 TO ESTABLISH DISK DATA SET 49
```

Figure 2(b). OVC-1: the CDC6600 version of the ANL code; this code is referred to as BISONMC in the text. The basis function documentation is the default; in actuality the integrals were computed from a contracted Gaussian basis with program PAINT.

modifications enable the printing of our output in a
neat, easily read format on narrow line devices. There
are also the additional advantages that the average
number of characters per line is increased (decreasing
the printing time) and that the narrow-width print-outs
can be filed in letter-sized folders to document the
run. Examples are displayed in Figures 2a, 2b, and 3a.
Figure 3b illustrates a 132 character/line format laid
out such that if it had been printed on a teletype, for
instance, the result would have still been a neat,
concise print-out with the right-hand columns vertically
aligned, but interleaved with the configurations on the
left.

Output--Summary Files. There are two reasons for
creating an output summary file. The first is to minimize the amount of time required to print essential information from a run. The second is to reduce the cumulative volume of the print-outs that one must store to
document the calculations. The capability to produce
summary files has been introduced into the Wright-
Patterson AFB CDC6600 versions of POLYINT, BISONMC,
and DASCI. An example for BISONMC is displayed in
Figure 3a. The output summary for the Wright-Patterson
version of BISONMC is dynamic in that, just as for the
punch file and except in the event of a termination by
the operator or a system crash, it is always produced,
and contains the orbitals at the last completed iteration. Program generated error messages are delivered
to the output summary file in a condensed manner.

Storage. There are many forms of storage available to the remote terminal user. First and foremost
among these forms are the tape and disk libraries of
the host computer. Possible storage media at the remote
terminal include paper tape, magnetic tape cassettes,
disks (standard and flexible), print-outs (teletype-
like and standard) and cards. None of these, except
for a very limited printer capacity, is indispensible
to a remote terminal for our type of computational
problem. Storage in the form of the traditional full
print-outs and card decks can be employed if operation
is from a batch terminal, or if there is someone at the
host site to mail output (perhaps on microfiche, if
available) if operation is from an interactive terminal.

Catalogued Procedures and JCCRGEN. The CDC job
control language is relatively easy to use - especially
for routine tasks. It does suffer from the fault that
a catalogued procedure cannot be modified at execution

```
OVC OUTPUT SUMMARY --- 05/J8/75  13. 0.2..--- JATA SET  1
OVC-I  MPAFB CDC6600 CANONICAL VERSION OF APRIL,1975
       NNO 1SI+ 8+3 P111
IO     PARAMETERS   0   0   0   4   +.0   0   0  -7   0   0   1   1   6
NCINPT PARAMETERS  0.0  19   0  1.0=-0+  1.0 -J9   5  0  0  0  0  2  2
ITERATION 2
ORBITALS  --- BLOCK 1
BF TYPE   ZETA     CENTR     C(1)      C(2)      C(3)       C(4)       C(5)       C(6)
 1 10  05909.4400    A      .00506    .00843    .59363    -.00750    .03804     .13638
 2 10  0   7.1327    A      .00686    .01155    .44633    -.01143    .00835     .13553
 3 10  0    .7000    A     -.01985   -.03355    .33070     .03464   -.02241    -.53553
 4 10  0    .2133    A     -.01029   -.00691   -.01432    -.01010    .34771    -.37743
 5 10  0  26.7860    A     -.00149   -.01034    .30005    -.01509   -.22332     .34310
 6 10  0    .5314    A     -.00110   -.00878   -.00077    -.03300   -.22020     .33743
 7 10  0    .1654    A      .00290    .00334    .00115    -.02303   -.02007    -.05932
 8 10  05909.4400    A      .00439    .59341   -.00073     .05108    .03173    -.01125
 9 10  0   7.1927    A      .00593    .45332   -.00103     .00950    .14000     .03253
10 10  0    .7000    A     -.01508   -.03712    .00293    -.05413   -.53031    -.03600
11 10  0    .2133    A     -.00718   -.02414    .00191     .07000   -.35145     .10401
12 10  0  26.7860    A      .00385    .01182   -.00153    -.04003    .03871     .01501
13 10  0    .5314    A      .00299    .01316   -.00314    -.02910    .13602     .02175
14 10  0    .1654    A      .00310    .00583   -.00132    -.02450    .03030    -.02373
15 10  07816.5400    A     -.59476   -.30335    .00033     .12254    .00127     .33133
16 10  0   9.5322    A     -.40700   -.00070   -.00050     .18142   -.00716     .03213
17 10  0    .9398    A      .01908    .03200    .00151    -.50794    .03913    -.03342
18 10  0    .2846    A      .01481    .03005    .00293    -.51357    .08093    -.05475
19 10  0  35.1832    A      .00760    .00754    .00078    -.05343    .06877     .30044
20 10  0    .7171    A      .00399    .01004    .30025    -.04315    .10624     .01120
21 10  0    .2137    A      .00469    .00591    .00042    -.06019    .02792     .01234
       WEIGHTED DEVIATION  1.6E-06  2.3E-06  1.9E-06  2.6E-06  2.9-06  3.6-06
BF TYPE   ZETA     CENTR     C(7)      C(8)      C(9)
 1 10  05909.4400    A     -.01866   -.07439    .00041
 2 10  0   7.1927    A     -.02247   -.11201    .00221
 3 10  0    .7000    A      .04354    .53220   -.04752
 4 10  0    .2133    A     -.02933    .80167   -.11504
 5 10  0  26.7860    A      .14980    .47382    .01794
 6 10  0    .5314    A      .11743    .65664   -.06533
 7 10  0    .1654    A      .07124    .21735   -.04043
 8 10  05909.4400    A     -.04052    .04431    .08220
 9 10  0   7.1927    A      .05948    .06815    .11523
10 10  0    .7000    A     -.17319   -.11329   -.53563
11 10  0    .2133    A     -.12886=1.07267  -.17826
12 10  0  26.7860    A     -.26662    .44308   -.45761
13 10  0    .5314    A     -.33642    .53817   -.52111
14 10  0    .1654    A      .04324    .29143   -.02203
15 10  07816.5400    A     -.00024    .00674   -.04597
16 10  0   9.5322    A     -.00226    .00859   -.06847
17 10  0    .9398    A      .01760   -.04369    .22734
18 10  0    .2846    A      .13020   -.07660    .01628
19 10  0  35.1832    A      .24611   -.14263   -.45864
20 10  0    .7171    A      .27532   -.05545   -.56136
21 10  0    .2137    A      .10873    .26672   -.13432
       WEIGHTED DEVIATION  2.4E-06  2.8E-06  5.3E-06
ORBITALS --- BLOCK 2
BF TYPE   ZETA     CENTR    C(10)     C(11)     C(12)      C(13)      C(14)      C(15)
22 20  0  26.7860    A     -.01440   -.26227    .34527    0.00000    0.00000    0.00000
23 20  0    .5314    A     -.00324   -.36493    .22244    0.00000    0.00000    0.00000
24 20  0    .1654    A      .05118   -.18597    .13488    0.00000    0.00000    0.00000
25 20  0  26.7860    A     -.18983   -.19657   -.33900    0.00000    0.00000    0.00000
26 20  0    .5314    A     -.27426   -.27809   -.43339    0.00000    0.00000    0.00000
27 20  0    .1654    A     -.10873   -.13113   -.14959    0.00000    0.00000    0.00000
28 20  0  35.1832    A     -.29682    .15899    .24340    0.00000    0.00000    0.00000
29 20  0    .7171    A     -.33463    .21021    .25702    0.00000    0.00000    0.00000
30 20  0    .2137    A     -.24905    .22070    .05984    0.00000    0.00000    0.00000
31 20  0  26.7860    A     0.00000   0.00000   0.00000   -.01440   -.26227    .34527
32 20  0    .5314    A     0.00000   0.00000   0.00000   -.00324   -.36493    .22244
33 20  0    .1654    A     0.00000   0.00000   0.00000    .05118   -.13597    .13403
34 20  0  26.7860    A     0.00000   0.00000   0.00000   -.18983   -.19657   -.33900
35 20  0    .5314    A     0.00000   0.00000   0.00000   -.27426   -.27809   -.43338
36 20  0    .1654    A     0.00000   0.00000   0.00000   -.10873   -.13113   -.14959
37 20  0  35.1832    A     0.00000   0.00000   0.00000   -.29682    .15899    .24340
38 20  0    .7171    A     0.00000   0.00000   0.00000   -.33463    .21021    .25702
39 20  0    .2137    A     0.00000   0.00000   0.00000   -.24905    .22070    .05984
       WEIGHTED DEVIATION  1.7E-06  1.1E-06  1.2E-06  1.7E-06  1.1E-06  1.2E-06
ENERGY-------------       CONFIGURATION EXPANSION COEFFICIENTS----
  -183.7673028162380       .95954  -.01425  -.01425  -.05764  -.00352
                          -.00352  -.12530  -.12530  -.04055  -.0+055
                          -.14024  -.04358  -.05706  -.07567  -.07567
                           .04769   .04769  -.04346  -.00741
  -183.0729003288399       .07730  -.00490  -.00049  -.00616  -.00589
                          -.30589   .32880   .32880  -.04274  -.04274
                          -.29544   .82718   .01762  -.00643  -.00643
                           .01020   .01020  -.04373  -.00053
       ORBITAL EXPANSION COEFFICIENTS CONVERGED TO  1.0E-04 IN ITERATION  2.
       OVC OUTPUT SUMMARY --- END
```

Figure 3(a). Output summary and further examples of 70-column output formatting. (a) OVC output summary for a 19-configuration MCSCF calculation for the $X^1\Sigma^+$ state of N_2O at its experimental equilibrium geometry with the Dunning (see Ref. 30) (4s3p)/[9s5p] contracted Gaussian basis set. The exponent of the first member of each contracted Gaussian is listed under the heading Zeta.

```
CONFIGURATION --- ENERGY/CI-COEFFS                                                                   -183.767303 -183.072500 -182.989553 -182.798993
 1. 2 2 2 2 2 2 2 0 0   2 2 0   2 2 0                                                                   .958537     .077302    -.031034     .000000
 2. 2 2 2 2 2 2 2 0 2   0 2 0   2 2 0                                                                  -.014246    -.000491    -.001325     .004571
 3. 2 2 2 2 2 2 2 0 2   2 2 0   0 2 0                                                                  -.014246    -.000491    -.001325    -.004571
 4. 2 2 2 2 2 2 0 0 2   2 2 0   2 2 0                                                                  -.057635    -.006162     .002595     .000000
 5. 2 2 2 2 2 2 2 0 2   2 0 0   2 2 0                                                                  -.003520    -.005890     .008477    -.013738
 6. 2 2 2 2 2 2 2 0 2   2 2 0   2 0 0                                                                  -.003520    -.005890     .008477     .013738
 7. 2 2 2 2 2 2 2 0 0   2 2 0   2 0 2                                                                  -.125297     .328802    -.481615    -.689701
 8. 2 2 2 2 2 2 2 0 0   2 0 2   2 2 0                                                                  -.125297     .328802    -.481615     .689701
 9. 2 2 2 2 2 2 2 0 0   2 2 0   0 2 2                                                                  -.040550    -.042736     .065756     .105050
10. 2 2 2 2 2 2 2 0 0   0 2 2   2 2 0                                                                  -.040550    -.042736     .065756    -.105050
11. 2 2 2 2 2 2 2 0 0   2 1 1   2 1 1   0 0 0 0 .5 .5 .5 .5 0.0 0.5 0.0     -.140235    -.295445     .458554    -.000000
12. 2 2 2 2 2 2 2 0 0   2 1 1   2 1 1   0 0 0 0 .5 .5 .5 .5 1.0 0.5 0.0     -.043583     .827185     .559060     .000000
13. 2 2 2 2 2 2 2 0 0   1 2 1   1 2 1   0 0 0 0 .5 .5 .5 .5 0.0 0.5 0.0     -.057356     .017621    -.053659    -.000000
14. 2 2 2 2 2 2 1 0 1   1 2 1   2 2 0   0 0 0 0 .5 .5 .5 .5 0.0 0.5 0.0     -.075674    -.006433     .002422     .017743
15. 2 2 2 2 2 2 1 0 1   2 2 0   1 2 1   0 0 0 0 .5 .5 .5 .5 0.0 0.5 0.0     -.075674    -.006433     .002422    -.017743
16. 2 2 2 2 2 2 1 1 0   2 2 0   2 1 1   0 0 0 0 .5 .5 .5 .5 .0 .5 .0         .047695     .010198    -.005189    -.112927
17. 2 2 2 2 2 2 1 1 0   2 1 1   2 2 0   0 0 0 0 .5 .5 .5 .5 0. .5 .0         .047695     .010198    -.005189     .112927
18. 2 2 2 2 2 2 2 0 0   2 0 2   2 0 2                                                                   .043458    -.043793     .023753    -.000000
19. 2 2 2 2 0 2 2 0 2   2 2 0   2 2 0                                                                  -.007406    -.000528     .000064     .000000

ITERATION TIMES (SECONDS) -- CP  64.67    IO  32.00
TOTAL MCSCF TIMES         -- CP 294.09    IO 156.00

RUN TERMINATION CONDITION  1 HAS ARISEN . . . I.E.
ORBITAL EXPANSION COEFFICIENTS CONVERGED TO  1.0E-04 IN ITERATION  2.
END OF DATA EXIT . . . FROM SUBROUTINE IO  IN OVLY(1,0)
```

Figure 3(b). Configurations for the run generating the aforementioned OVC output summary. The layout of this line-printer page is such that if printed on a teletype the configuration coefficients for the various CI roots in the columns on the right will come out neatly interleaved with their respective configurations on the left.

time by parameters entered on the control card which
calls it into execution. Nonetheless, it is possible
to run a single program with only a few, simple job
control statements. Initially we did so. Later it
became necessary to use even more complicated job control card records (JCCR's) to solve the problems involved in executing multiple job steps. When the capacity to catalog these procedures became available, it
was used to reduce input job formulation errors and to
minimize the knowledge of the CDC job control language
required in the construction of an input deck. Figure
6b is an example of one such catalogued procedure.

Now we have gone one step further. We have written a program, JCCRGEN, in FORTRAN IV to perform the
task of substituting the proper CDC control card sequences for each of a limited number of run-specifying
commands. These commands are entered by the user at
the head of the job in record two. This program is
keyed to our immediate needs, but it can easily be expanded to handle other codes and file manipulation
tasks. The role that this program performs on the CDC
system is, in essence, available implicity in IBM job
control language via the capability one has there to
write catalogued procedures which can be modified at
each execution by means of parameters entered on the
EXEC card. Our job control card record generation
(JCCRGEN) program is invoked by the use of a job deck
set-up as shown in Figure 4. The commands are displayed
in Figure 5. Figure 6a shows how simple the overhead
is in the application of JCCRGEN to the task of the construction of the job control card record (JCCR) shown
in Figure 6b. Figure 7 displays a simple job that one
might run to retrieve a file from an archive tape on
the host computer and place a copy of it on a disk file
for subsequent use in constructing a new job or for
perusal purposes.

Batch Terminals. We have had no direct experience
going into WPAFB with remote batch terminals. However,
there are, of course, individuals who have successively
employed them for ab initio calculations. Their advantage over remote interactive terminals is that they
permit one to operate from cards and with punched and
printed output much as if he were "on-site". This
latter is particularly true if the transmission rate is
4.8 K/B or higher. Their disadvantage, as can be seen
from Table I, is that their purchase price is many
times that of a teletype or other interactive terminal.

```
JOB CARD
ATTACH, JCCRGEN.
JCCRGEN.
CCLINK, JCCR.
*EOR
  COMMANDS FOR JCCRGEN SEPARATED BY ONE OR MORE SPACES
*EOR
  INPUT DATA SETS NOT MADE AVAILABLE BY THE JCCRGEN
  COMMANDS IN THE HIERARCHIAL SEQUENCE BISON OR PAINT
  BEFORE BISONMC BEFORE DASCI.  SEPARATE THE DATA SETS
  WITH AN END-OF-RECORD (*EOR,LEFT-JUSTIFIED).
=
```

Figure 4. General form of a remote interactive terminal input job that uses JCCRGEN via CDC INTERCOM. For a remote batch terminal a card with a column 1 multi-punch 7/8/9 replaces the *EOR; a column 1 multi-punch 6/7/8/9 replaces the =.

```
TERMINAL DESIGNATION--INTERACTIVE OR BATCH.  MUST BE FIRST COMMAND
INTERTERM (IT)     This is the default.
BATCHTERM (BT)

CODE RETRIEVAL AND EXECUTION COMMANDS.     FORM:    COMMAND, OPTIONS.
BISON     (or BI)             BIMC
BISONMC   (or MC or OVC)      PAMC
DASCI     (or CI)             PAMCCI
PAINT     (or PA)             MCCI       (or OVCCI)

INTEGRAL STORAGE OPTION

BISON     TAPE,file label,tape #
PAINT   = DISK,pf lable,cycle #,file #
          TAPE,tape #,file #
INTEGRAL RETRIEVAL COMMAND
INTEGRALS_TAPE,file label,tape #
(or INT)  DISK,pf label,cycle #,file #
OUTPUT FILE STORAGE COMMANDS AND COMMAND OPTIONS
BASIC FORM:    COMMAND=TAPE,file label,tape #
                      DISK,pf label,cycle #,file #

PRINTSAVE    (or PRS)       CIPUSAVE       (or CIPUS)
INDECKSAVE   (or IDS)       EXITDECKSAVE   (or EDS)
OUTDECKSAVE  (or ODS)       EXITPRINTSAVE  (or EPRS)
MCPUSAVE     (or MCPUS)     PRINTPRINT     (or PRPR)
                            PUNCHPUNCH     (or PUPU)

FILE RETRIEVAL COMMANDS AND COMMAND OPTIONS
BASIC FORM:    COMMAND=TAPE,file label,tape #
                      DISK,pf label,cycle #,file #

PAINPUT   (or PAIN)       BIINPUT     (or BIIN)
MCINPUT   (or MCIN)       TAPETODISK  (or TTD)
CIINPUT   (or CIIN)       DISKTOTAPE  (or DDT)

COMMENTS

"label" is 1-17 alphanumeric characters for tape, 1-40 for
disk.  "file #" need not be given if it is 1.
```

Figure 5. Commands for JCCRGEN

6(a)

```
DHACS,CM225000,T1500,IO1000,MT1. R750825,HOPPER,ARL/LJ,54869.
ATTACH,JCCRGEN.
JCCRGEN.
CCLINK,JCCR.
*EOR
INT=TAPE,NNOB43P111,L00001
OVC,MCPUS=TAPE,NNOB43P111X1SIP,L00002
CI,CIIN=TAPE,X1SIPCONFIGS,L00003
PRS=TAPE,NNOB43P111X1SIPPR,L00004
EPRS=TAPE,NNOEXITSAVE,L=00005
EDS=DISK,NNOEXITSAVE,6
*EOR
---input data set for BISONMC---
=
```

Figure 6(a). *JCCRGEN Example 1. (a) Depiction of an input job formulated in the EDITOR mode of CDC6600 INTERCOM which uses JCCRGEN commands to (1) retrieve a stored integral set from tape, (2) perform an MCSCF calculation, (3) archive the converged orbitals on tape in the format of a BISONMC input deck, (4) formulate an input data set for the DASCI program from the final orbitals from the BISONMC run and an archived CI configuration list, (5) execute a DASCI run, (6) augment the printed output file with an 80–80 listing of the entire input job deck and with a copy of the output summary file (TAPE57), (7) archive a copy of the augmented printed output file on tape, and, (8) in the event of the encounter of an abnormal run termination condition, archive a copy of the output file on tape and create and store on disk a completed input job deck equivalent in all detail to the original except that the orbitals contained are those from the last successfully completed MCSCF iteration. Job control passes to the file JCCR upon encounter of the command CCLINK,JCCR.*

PAUSE. THIS IS A REMOTE TERMINAL JOB (E.G.TTY).
PAUSE. PLEASE GET MY TAPES FROM THE LIBRARY.
VSN,MFN1=L00001,MFN2=L00002,MFN3=L00003,MFN4=L00004,MFN5=L00005.
REQUEST,MFN1,MF,NORING.
LABEL,T43,M=MFN1,R,L=NNOB43P111.
COPYBF,T43,TAPE43.
UNLOAD,MFN1.
REWIND,TAPE43.
LIMIT,7777.
ATTACH,BISONMC,LOADMODS,CY=3,MR=1.
BISONMC.
REQUEST,MFN2,MF,RING.
LABEL,ODS,M=MFN2,W,L=NNOB43P111X1SIP.
REWIND,PUNCH.
COPYBR,PUNCH,ODS.
UNLOAD,MF2.
REWIND,TAPE61.
COPYBR,TAPE61,CIDECK.
REQUEST,MFN3,MF,NORING.
LABEL,CIIN,M=MFN3,R,L=X1SIPCONFIGS.
COPYBR,CIIN,CIDECK.
UNLOAD,MFN3.
REWIND,CIDECK.
REWIND,TAPE43.
ATTACH,DASCI,LOADMODS,CY=4,MR=1.
DASCI,CIDECK.
REWIND,INPUT.
COPYBF,INPUT,OUTPUT.
REWIND,TAPE57.
COPYBF,TAPE57,OUTPUT.
REWIND,OUTPUT.
REQUEST,MFN4,MF,RING.
LABEL,PRS,M=MFN4,W,L=NNOB43P111X1SIPPR.
COPY,OUTPUT,PRS.
UNLOAD,MFN4.
EXIT.
REWIND,INPUT.
COPYBF.
REWIND,TAPE57.
COPYBF,TAPE57,OUTPUT.
REQUEST,MFN5,MF,RING.
LABEL,EPRS,M=MFN5,W,L=NNOEXITSAVE.
REWIND,OUTPUT.
COPY,OUTPUT,EPRS.
UNLOAD,MFN5.
REQUEST,ORBSAVE,*PR.
REWIND,INPUT,PUNCH.
COPYBR,INPUT,ORBSAVE,2,0.
SKIPF,INPUT,1,0.
COPYBR,PUNCH,ORBSAVE.
COPYBF,INPUT,ORBSAVE.
CATALOG,ORBSAVE,NNOEXITSAVE,CY=6,RP=999.

Figure 6(b). The job control card record (JCCR) file generated by the JCCRGEN commands in Figure 6a

Summary and Future Plans

Summary. We have demonstrated that it is possible to execute state-of-the-art ab initio quantum molecular calculations remotely with an interactive terminal (37). Participants in our computational chemistry project have used teletypes to perform OVC-MCSCF calculations at Wright-Patterson Air Force Base, Ohio, from Argonne, Illinois, and Washington, D. C., and equations-of-motion calculations from Salt Lake City, Utah. It has been possible to sustain a moderate production rate in this manner. We have also demonstrated that it is possible to maintain and update major computer codes from such a simple remote terminal.

Future Plans -- To Ease Remote Operations. There are several things that we hope to do to ease remote operations on our project. One is the use of a remote batch terminal. More importantly, it may soon be possible to use network connections to reduce the problems associated with the communication link--speed and cost. Most networks now extant operate at 4800-50,000B, compared to our 110-300B teletype transmission rate (35). In our particular case the call to WPAFB from ANL could go over the ARPANET at 50 K/B in about a year from this writing, when the computer centers at both sites have become fully operational nodes on this network. See Figure 8. It is important to note that, so far as our project is concerned, the communication costs will be negligible since the phone call would be to a network node within a local call radius.

Another possible future solution to the communication speed problem would be to engage the services of one of the commercial companies now going into the sole business of establishing and providing high speed (50, 230, 1300 K/B), low undetected error rate (1 in 10^{12} bits), data transmission conduits over special ground lines and by microwave and satellite transmission (35).

The expansion of networks and the increased availability of high speed data communication services over the next few years should combine to increase the speed while decreasing the cost. The limit to the maximum usable transmission rate will then become the terminal and terminal-to-node transmission speed rather than the capacity and quality of the long distance transmission link. It is, thus, appropriate to anticipate the development of high-speed radio-terminals to complete the solution to the present day data communication bottleneck to our type of data processing application.

11. HOPPER ET AL.   *Calculations via Remote Terminal*   175

```
DHACS2,CM10000,T10,IO10,MT1.R750825,HOPPER,ARL/LJ,54869.
ATTACH,JCCRGEN.
JCCRGEN.
CCLINK,JCCR.
*EOR
TTD   TAPE,NNOB43P111X1SIP,L00002
DISK,DHWORKFILE,1
=
```

Figure 7(a). JCCRGEN—Example 2. (a) Depiction of an input job, formulated on an interactive terminal using the EDITOR mode of CDC6600 INTERCOM, which employs JCCRGEN commands to retrieve an archived file from tape and produce a copy on disk. The latter is then available for perusal or editing to produce an input deck via the EDITOR mode.

```
PAUSE. THIS JOB IS FROM A REMOTE TERMINAL (E.G.TTY).
PAUSE. PLEASE GET MY TAPES FROM THE LIBRARY.
REQUEST,MFN1,MF,VSN=L00002,NORING.
LABEL,TAPE,R,L=NNOB43P111X1SIP.
REQUEST,DISK,*PF.
COPYBF,TAPE,DISK.
CATALOG,DISK,DHWORKFILE,CY=1,RP=999.
```

Figure 7(b). The JCCR file generated by the JCCRGEN commands in Figure 7a.

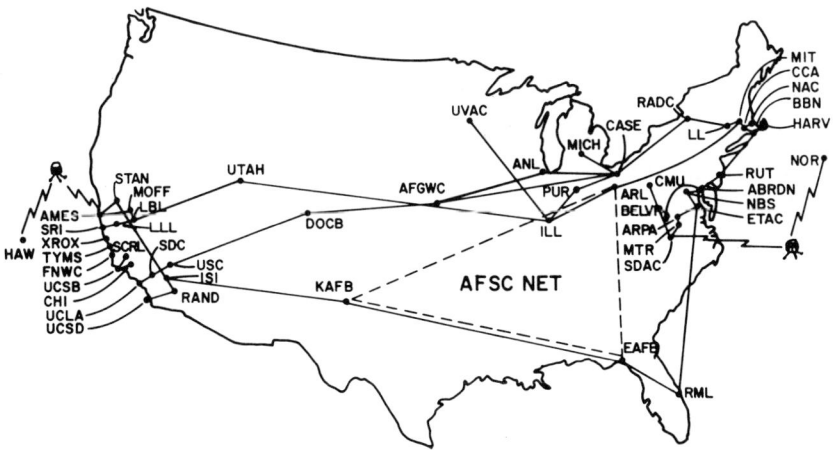

Figure 8. ARPA network and the proposed "piggy-back" AFSC test network.

The future installation and maintenance of codes at WPAFB on this project might be performed entirely remotely. There are two ways we anticipate that we might go about this: (a) with a remote terminal equipped with a card reader and an initial source deck on physical cards, or (b) with an interactive terminal equipped with, e.g., tape cassettes and an initial recording (over perhaps several cassettes) of the program.* It is pertinent in this regard to note that 120,000-240,000 characters may be contained on one tape cassette, depending on the recording mode.** Thus, the entire BISONMC source code could be contained on six or three cassettes, again depending on the recording mode.

Future Plans--Air Force Systems Command Network. The development of the ASD Computer Center at Wright-Patterson AFB is, of course, continuing, in order to establish capabilities beyond those already described. During the past two years, the Air Force Systems Command has been considering procedures to link the computers at its various technical centers located throughout the country. The major motivation is to permit resource sharing between equivalent computers. Several studies of the best methods to effect this linking were commissioned by SADPR-85 (the Support of Air Force Automatic Digital Processing Requirements Mission Analysis for 1985). In December 1974, it was proposed that a test network (AFSCNET), consisting of the computer centers at Eglin, Kirtland, and Wright-Patterson Air Force Bases be set up to ride piggy-back on the ARPA network (ARPANET). See Figure 8.

The ARPA network was developed by the Bolt-Baranek-Newman Co. under a contract from the Advanced Research Projects Agency monitored by the Air Force Systems Command (6, 8, 9, 38-41). From the first four nodes (U.C.-Santa Barbara, University of Utah, Standford Research Inst., UCLA) in about 1968-9 the network has expanded rapidly so that by 1973-4 about 47-48 university and

---
*The ANL-WPAFB long distance phone bill for a 5000 card source deck would be on the order of $10 at 2400B or $5 at 4800B, but prohibitive at lower speeds. A call on the ARPANET would be local and terminal-limited in speed.
**A "Philips-type" cassette which contains 300 feet of 0.15-inch-wide magnetic tape with serial data recording at 47 characters/inch and inter-record gaps of 2.2 inches has a storage capacity of 483 256-character records for a total of 123,648 characters (35). Two-track recording allows 246,296 characters/cassette.

government sites were interconnected. See Figure 8. The ARPANET was established as an experimental, not an operational, network. However, with its transition from ARPA support, it seems to be becoming more operationally oriented. The proposed AFSCNET is an example.

Each node on the proposed AFSCNET currently possesses a pair of CDC6600's or equivalent machines (i.e. CYBER73). The plan is to employ a PDP-11 to control the flow of jobs between the network and the on-site computers on the one hand, and between the hard-wired batch terminals (CDC1700's) and the on-site computers or the network on the other. See Figure 9. It is expected that an IBM 370 can be included by direct coupling to the CDC6600 as well as to the PDP11. The acquisition of an IBM 370/155 is currently under study. Interactive terminal access at 110, 300, and 1200B is to be provided to all on-site computers and to remote users through the network. Remote batch terminals operating at 2 K/B and 4.8 K/B are expected to have access to a CDC6600 and through it to other computers on-site and on AFSCNET.

In about one year enough components of the system should be completed to enable sharing of work-load overflow among the CDC6600's at the three Air Force bases in the test network. By that time too, Argonne National Laboratory will be in operation on the ARPANET so that it might be -- at least in theory -- possible to perform calculations on the Wright-Patterson computers by complete remote operation from Argonne. Few site visits should then be required and this could lead to the establishment of a network in computational chemistry of enormous power and versatility.

Concluding Remark. It is realistic, we think, to expect that the now-developing expertise in the remote execution of production runs of sophisticated codes will make this activity common place. Much work remains to be done, but it seems likely that this development could be highly useful to the researcher who may only upon occasion need to execute a set of quantum-chemical, scattering, kinetic, or other type of calculations in connection with his major research activities. This development will make the sophisticated instruments of computational chemistry accessible to the average laboratory or office and to the finger-tips of virtually every chemist.

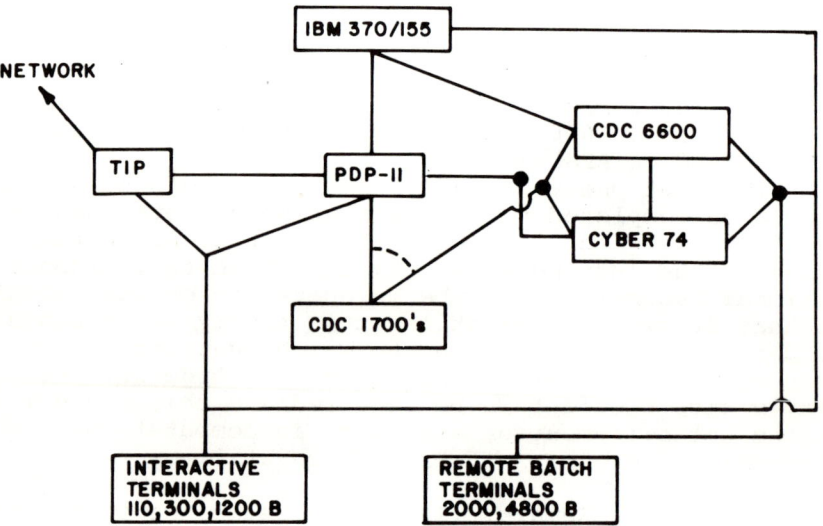

Figure 9. Example diagram for the proposed AFSC network test sites at Eglin, Kirtland, and Wright-Patterson Air Force Base.

Acknowledgements

We thank Mr. Demetrius C. Zonars of the Aeronautical Systems Division Computer Center, Air Force Systems Command, Wright-Patterson Air Force Base, for providing us with the cronology of computational equipment and contractor usage. Dr. Paul Nicolai of the Mathematical Research Laboratory, Aerospace Research Laboratory, Air Force Systems Command, Wright-Patterson Air Force Base, provided the information on the Wright-Patterson role in the ARPANET and the plans for an AFSC network. Dr. Paul Messina of the Applied Mathematics Division, Argonne National Laboratory, provided the information concerning Argonne's participation in the ARPANET. Dr. Nicolai and the open shop personnel at the ASD Computer Center, Mr. Jim Bittle and Mr. Jim Hudson in particular, have aided our remote usage effort tremendously in discussion of the details of implementing our IBM codes on the CDC6600 and in obtaining solutions to routine operating problems. Much of the preliminary work in establishing our remote usage procedures was performed by Dr. Richard P. Hosteny. Dr. Richard J. Blint aided in the establishment of the Utah connection. We thank Dr. Bob Euwema of the theoretical solid-state group, Aerospace Research Laboratories, Air Force Systems Command, Wright-Patterson AFB, for permitting us to quote from his letter from Dr. S. Huzinaga of the University of Alberta, Canada and for answering our numerous questions about running sophisticated codes on the CDC6600.

Abstract

A historical perspective of quantum chemical calculations, starting with the master integrals-SCF codes constructed and extensively employed on-site at Wright-Patterson AFB on the Univac 1103 during the approximate period 1956-1961 and evolving to our current remote usage, is sketched. These two decades of experience at Wright-Patterson provide a currently developing example of computer resource sharing directly related to chemistry.

Procedures have been devised for the execution of ab initio quantum mechanical computations from a remote interactive terminal. Most of the procedural and software tactics apply to the execution of such calculations from a remote batch terminal as well. While the optimun - and most expensive - situation is to have both interactive and batch terminals available along with a high speed data link to a computer, a simple teletype utilizing regular phone lines is quite sufficient to engage in this activity if the user is properly organized on the host computer. Our experience with such remote usage is discussed.

Literature Cited

1. Karo, A., Krauss, M., and Wahl, A. C., Int. J. Quantum Chem. (1973) 75, 143.
2. Schaefer, H. F., "The Electronic Structure of Atoms and Molecules; A Survey of Rigorous Quantum Mechanical Results," Addison-Wesley, Reading, Mass., 1972.
3. For example, the Quantum Chemistry Program Exchange, University of Indiana, Bloomington, Indiana, now has broaden its scope to the collection and distribution of computer codes in all areas of computational chemistry. QCPE is beginning to perform some service calculations as well.
4. Hall, G. G., Chem. Soc. Rev. (1973) 2, 21.
5. The term "interactive" (or, equivalent, "time-sharing") terminal is taken to mean asynchronous transmission at 110B, 300B, 1200B, etc. The term "batch terminal" is taken here to mean a terminal capable of synchronous transmission at 2 K/B or 4.8 K/B over normal voice grade phone lines. The adjective "remote" is taken to imply access from off-site throughout this paper. The "host computer" is understood to be off the premises of the user's institution.
6. Peterson, J. J. and Veit, S. A., "Survey of Computer Networks," The MITRE Corp., September, 1971. Available from the National Technical Information Service, U. S. Department of Commerce, Springfield, Virginia 22151.
7. Greenberger, M., Aronofsky, J., McKenney, J. L., and Massy, W. F., "Networks for Research and Education - Sharing Computer and Information Resources Nationwide," MIT Press, Cambridge, Mass., 1974.
8. Bouknight, W. J., Grossman, G. R., and Grothe, D. M., "The ARPA Network Terminal System - A New Approach to Network Access," Proc. Data Communications '73, Nov. 1973.
9. Sher, M. S., "A Case Study in Networking", Datamation (1974) March, 56-59.
10. This history is constructed mainly from the "Introduction to ASD Computer Center", Wright-Patterson Air Force Base, Ohio, May, 1974 and from material furnished by Mr. D. Zonars and Dr. P. Nicolai at Wright-Patterson.

11. Readers Digest 1975 Almanac and Yearbook, The Reader's Digest Association, Inc., Pleasantville, New York, 1975.
12. Ransil, B. J., Revs. Mod. Phys. (1960) $\underline{32}$, 239.
13. Ransil, B. J., Revs. Mod. Phys. (1960) $\underline{32}$, 245.
14. McLean, A. D., Weiss, A., and Yoshimine, M., Revs. Mod. Phys. (1960) $\underline{32}$, 211.
15. Kolos, W. and Roothaan, C. C. J., Revs. Mod. Phys. (1960) $\underline{32}$, 219.
16. Kolos, W. and Roothaan, C. C. J., Revs. Mod. Phys. (1960) $\underline{32}$, 219.
17. Richardson, J. W., Revs. Mod. Phys. (1960) $\underline{32}$, 461.
18. Kolos, W., Roothaan, C. C. J., and Sack, R. A. Revs. Mod. Phys. (1960) $\underline{32}$, 178.
19. Roothaan, C. C. J., Revs. Mod. Phys. (1960) $\underline{32}$, 179.
20. Roothaan, C. C. J., and Weiss, A. W., Revs. Mod. Phys. (1960) 32, 194.
21. Fraga, S. and Mulliken, R. S., Revs. Mod. Phys. (1960) $\underline{32}$, 254.
22. Fröman, A., Revs. Mod. Phys. (1960) $\underline{32}$, 317.
23. Löwdin, P. O., Revs. Mod. Phys. (1960) $\underline{32}$, 328.
24. Wahl, A. C., Bertoncini, P., Kaiser, K., and Land, R. H., "BISON - A Fortran Computer System for the Calculation of Analytic Self-Consistent-Field Wavefunctions, Properties, and Charge Densities for Diatomic Molecules: Part 1. User's Manual and General Program Description," Argonne National Laboratory Report ANL-7271 (January, 1968). For availability see Ref. 6.
25. Das, G. and Wahl, A. C., "BISON-MC: A FORTRAN Computing System for Multiconfigurational Self-Consistent-Field (MCSCF) Calculations on Atoms, Diatoms, and Polyatoms," Argonne National Laboratory Report ANL-7955 (July, 1972). For availability see Ref. 6.
26. DASCI is a small (100) configuration interaction program written by G. Das, Argonne National Laboratory for use with BISON-MC.
27. Neuman, D., et al., "POLYATOM (Version 2)," Bell Labs. Rep.; PA300 was modified and linked to BISON-MC by A. Hinds, Argonne National Laboratory; we refer to this modified version, which generates canonical lists of integrals and no labels, as POLYINT.
28. Das, G. and Wahl, A. C., J. Chem. Phys. (1972) $\underline{56}$, 1769. See also Ref. 1.
29. Hosteny, R. P., Hinds, A. R., Wahl, A. C., and Krauss, M., Chem. Phys. Letters (1973) $\underline{23}$, 9.

30. Hopper, D. G., Chem. Phys. Letters (1975) 31, 446.
31. Fortune, P. J., Rosenberg, B. J., Das, G., and Wahl, A. C., Abstracts of the Eighth Midwest Theoretical Chemistry Conference, University of Wisconsin, Madison (May 1975) 8, 26.
32. See, e.g., Smith, W. D., Chen, T. T., and Simons, J., J. Chem. Phys. (1974) 61, 2670.
33. Krauss, M., Hopper, D. G., Hosteny, R. P., Fortune, P. J., Wahl, A. C., and Tiernan, T. O., "Potential Energy Surfaces for Atmospheric Triatomic Molecules. I. Literature Review," ARL Tech. Rep., June, 1975. For availability see Ref. 6.
34. Fortune, P. J., Hopper, D. G., Rosenberg, B. J., England, W. B., Gillespie, G., Hosteny, R. P., Das, G., Wahl, A. C., and Tiernan, T. O., "Potential Energy Surfaces of Atmospheric Triatomic Molecules. II. Results of SCF and Preliminary OVC-MCSCF-CI Calculations," ARL Tech. Rep., June 1975, and references therein. Availability — Ref. 6.
35. "Datapro 70, the EDP Buyer's Bible," Datapro Research Corporation, Morris Town, N. J., 1972, and updates through 1975.
36. Carter, C., "Guide to Reference Sources in the Computer Sciences," MacMillan Information, MacMillan Publishing Co., N. Y., 1974.
37. Harris, F. E. and Ault, R. H., "Abstracts of the 169th ACS National Meeting," paper COMP30, Amer. Chem. Soc., Washington, 1975, have discussed, for another type of computer, software ideas similar to some of ours.
38. Heart, F. E., Kahn, R. E., Ornstein, S. M., and Crowther, W. R. and Walden, D. C., Proc. SJCC '70 (1970), 551. Description of the ARPANET IMP (Interface Message Processor).
39. McQuillan, J. M., Crowther, W. R., Cosell, B. P., Walden, D. C., and Heart, F. E., Proc. FJCC '72 (1972), 741. ARPA improvements.
40. Ornstein, S. M., Heart, F. E., Crowther, W. R., Rising, H. K., Russell, S. B., and Michel, A., Proc. SJCC '72 (1972), 243. Description of the ARPANET TIP - terminal IMP.
41. McKenzie, A. A., Cosell, B. P., McQuillan, J. M., and Thorpe, M. J., Proc. ICCC '72 (1972), 185. Description of an ARPANET control center.

12

Computer Identification and Interpretation of Unknown Mass Spectra Utilizing a Computer Network System

R. VENKATARAGHAVAN, GAIL M. PESYNA, and F. W. McLAFFERTY

Department of Chemistry, Cornell University, Ithaca, N. Y. 14850

Mass spectrometry/computer systems in routine use in many research laboratories are capable of producing a complete mass spectrum (unit resolution) every few seconds (1, 2). The unique applicability of mass spectrometry to nanogram samples, and the ability to obtain spectra directly on components of complex mixtures separated by a gas (3) or liquid chromatograph (4) have tremendously increased the number of spectra taken for the purpose of compound identification. These capabilities have provided unique solutions to research and control problems in a wide variety of fields, including environmental pollution, metabolism studies, medical diagnoses, insect pheromones, forensic analyses, military detection systems, and conversion of coal or shale to liquid fuels.

The importance of these applications has motivated a large increase in basic research in organic mass spectrometry, greatly expanding our knowledge of mass spectral fragmentation behavior. Unfortunately, the quantity (and quality) of chemists trained in the interpretation of mass spectra has not increased as rapidly, and thus the interpretation process is becoming an increasingly serious bottleneck in proper utilization of this technique.

The modern computer is an obvious possibility to alleviate these problems, and a number of computer systems for identification and interpretation have been proposed (1, 2, 5-13). If a reference mass spectrum of the unknown compound is in the data file, computer matching programs can be used for its retrieval. Although such retrieval is relatively efficient, the present mass spectral reference file (14, 15) of 30,000 different compounds represents a very small fraction of the possible organic compounds, so that interpretation is required whenever a poor match is obtained. Thus both "retrieval" and "interpretive" systems are necessary to solve problems in most important areas. These systems have been reviewed in detail recently (1, 2, 5); here we will only describe the Cornell systems which are now available over a computer networking system (TYMNET).

PBM. A "Probability Based Matching" system for the retrieval of unknown mass spectra has recently been described (12, 13). Research in the area of document and information retrieval has firmly established that system efficiency is increased by the proper weighting of the relative importance of items used for identifying each member of a library (16). For PBM the m/e values of the peaks are weighted according to their uniqueness in the file, and the abundance values are weighted according to a log normal distribution (17). The use of these values is based upon the "General Rule of Multiplication" of probability theory; thus if peaks with masses m_1 and m_2 having intensities i_1 and i_2 occur in mass spectra with probabilities p_1 and p_2, the probability that both occur at random in an unknown spectrum is p_1 times p_2. If this product is small, it is much more likely that the presence of peaks m_1 and m_2 is due to the identity of the unknown spectrum and the compared reference spectrum in which both occur with the intensities i_1 and i_2. The low value of this probability provides a confidence that this identification is correct, which is measured by a "confidence value, K". This measure, as well as all the individual probabilities, is expressed as the corresponding base two logarithm for convenience of calculation; inverse probabilities are also used to simplify the calculations and to produce a final result which is a direct measure of "confidence." In this reverse search, there is computed for each reference spectrum matched against the unknown a confidence value, "K", equal to the sum of the individual K_j values calculated for each peak in the unknown whose intensity agrees within a predetermined range to that of the corresponding peak in the reference spectrum. K_j combines four terms,

$$K_j = U_j + A_j + W_j - D$$

where U is the contribution to the probability of the "uniqueness" of the m/e value of the peak; A is the contribution to the probability of the abundance value of the peak as it appears in the reference spectrum; W, the "window factor", is a measure of the agreement required between the abundance of the peak in the reference and in the unknown; and D, the "dilution factor" for mixture spectra, is a measure of the overall reduction of peak intensities in the unknown due to the presence of other components (if the unknown spectrum is of a pure compound, D = 0). The system is described in detail elsewhere (13).

Extensive statistical studies have shown (13) that the precision/recall performance of the system is substantially better than others which employ no or limited weighting. The "reverse search" feature is especially valuable with mixtures, as illustrated by results from the mass spectrum of a mixture shown in Table I. The mixture contained amobarbital (5-ethyl-5-isoamyl-

Table I. PBM Results for an "Unknown" Mass Spectrum of 30% Amobarbital, 30% Hexobarbital, and 40% Nicotine[a]

Compound	Confidence Value K^b	ΔK	% Contamination	% Component
5-cyclohexenyl-1,5-dimethylbarbituric acid (hexobarbital, cyclonal)	109+, 82*+ 82*+, 75+ 60+, 50***+ 45***+	10, 37 37, 44 59, 69 74	59, 59 59, 72 80, 79 79	74, 55 55, 32 37, 44 44
nicotine	77+, 77+ 72*+, 61** 57+, 54**	42, 42 47, 58 62, 65	74, 74 74, 74 74, 74	99, 95 70, 70 69, 100
5-ethyl-5-isoamyl-barbituric acid (amobarbital, amytal)	76+, 66 66, 64** 56, 48*	43, 53 53, 55 63, 71	62, 62 62, 62 62, 62	73, 60 60, 58 61, 51

[a]Compounds selected of highest K values from a data base of 35,828 spectra.
[b]Multiple values represent different spectra of the same compound; the asterisks indicate the number of "flagged" peaks (see text) omitted in calculating that K value, and the plus sign indicates that the molecular ion of the reference was found in the unknown and used in the K calculation.

barbituric acid), hexobarbital (5-cyclohexenyl-1,5-dimethylbarbituric acid), and nicotine. PBM has successfully retrieved a relatively large number of spectra of the correct compounds, measured under a wide variety of conditions, without retrieving a single incorrect compound of comparable K value. The mass spectrum of pentobarbital, the sec-amyl isomer of the first compound, is closely similar to that of amobarbital, so that the selectivity exhibited by PBM in this case, despite the presence of a second barbiturate, is gratifying.

STIRS. For the "Self-Training Interpretive and Retrieval System" (11) a number of classes of mass spectral data known to have high structural significance, such as characteristic ions, series of ions, and masses of neutrals lost, are identified; for each class the computer matches the data of the unknown mass spectrum against the corresponding data of all reference spectra. In each data class the reference compounds whose spectra have the highest "match-factor" (MF) values are examined by the chemist for any common structural features, with a high frequency of occurrence indicating a high probability that the structural feature is present in the unknown. In a recent modification (18) the 15 selected compounds of highest MF value are examined instead by the computer for the presence of specific substructural groups to provide a statistical evaluation of the probability of the presence of each group in the unknown compound. At present STIRS is able to predict the presence of 179 substructures at the 98 percent confidence level with an average recall of 49 percent (i.e., using criteria in which STIRS is wrong only once in 50 times, the substructure can be identified in half of the compounds in which it is actually present). Because of the nature of mass spectra we do not have STIRS attempt to predict the absence of a substructure; the influence of a particular substructure on the mass spectral behavior can be greatly reduced by the presence in the molecule of another substructure which more strongly directs the fragmentation.

Structural data for all compounds in the reference file have been coded in Wiswesser Line Notation (WLN), which is a linear representation of the compound's structure requiring a relatively small volume of computer storage. In WLN, symbols such as 1, R, M, Z, V are used to represent individual chemical units such as $-CH_2-$, C_6H_5, $>NH$, $-NH_2$, $\overset{O}{\underset{-C-}{\parallel}}$ respectively; connection transfers are used to show breaks from linear representation, ring units to show the presence of rings in the structure, and ring fusion units to show ring interrelations in a multicyclic system. In comparison to other modes of structure representation, WLN has several special advantages for the interpretation of mass spectral data. In an acyclic system the linear notation is useful in identifying fragmentations resulting from simple cleavages, the chemical units of WLN often being directly related to masses and mass differences in the spectrum.

The presence of certain chemical units such as V (carbonyl) and Z (primary amino) can exert a strong influence on the fragmentation behavior of the molecule and can thus be readily used for the identification and assignment of elemental compositions to different ions. Deviations from structural linearity are designated clearly in the notation by connection transfers, and can thus be recognized by the computer system to relate points of branching and substitution in a molecule to mass spectral cleavages commonly triggered by such structural features. Complex cyclic molecules are among the most difficult types of structures for mass spectral interpretation; the ability of WLN to represent the relationships of different rings and substituents within the structure often makes it possible for the computer to recognize major fragmentations and to develop spectra-structure correlations. As an example (11), running the spectrum of cholesterol as an "unknown", STIRS for match factor 5 selected neutral losses of 18, 0 (M^+), 33, 15, 17, and 61; the six compounds found with the highest MF5 values are the steroidal derivatives allopregnanol-3α-one-20, pregnenolone alcohol, pregnenolone, 7β-hydroxycholestanyl 3β-acetate, 16α-methylpregnenolone (a large number of of other oxygenated carbocyclic compounds are present in the data base). Although the loss of these simple neutral species from cholesterol is quite <u>consistent</u> with present knowledge of the mass spectral behavior of such molecules, it is doubtful that these losses were known to be <u>characteristic</u> of such molecules.

The "Artificial Intelligence" (9) method uses the computer to apply human mass spectral knowledge to predict spectra of isomeric possibilities. As far as possible all of the known mass spectral fragmentation behavior is programmed into the computer, and then the mass spectra of feasible isomers (the possibilities are generated by the DENDRAL algorithm) are predicted and compared to the unknown mass spectrum. This demands that the unknown compound be in the rather narrow class for which the program is written and that its elemental composition has been established. An Artificial Intelligence program for estrogens (10) appears to be the only one that has been tested extensively on true unknowns. STIRS is complementary to the "Artifical Intelligence" technique in that it can be applied to the spectra of total unknowns (such as pollutants, insect secretions, and abnormal urinary constituents) to obtain partial structural information. For example, STIRS in general can easily identify estrogens and often some of the substitutents thereon, but the Artificial Intelligence method is superior for properly placing the substituents and elucidating other structural details. STIRS utilizes directly the information of all available reference spectra without prior spectra-structure correlation, "training" itself separately for each submitted unknown spectrum; thus the only special preparation necessary to make STIRS sensitive for estrogens is to make sure that there are representative mass spectra of estrogens in the reference data base.

Chemists sometimes forget the potential complexity of the problem of determining the exact molecular structure of a new compound of a molecular weight of even a few hundred. Although a relatively small amount of information usually can greatly reduce the millions of possibilities (billions if less common elements are included), the effort required to narrow the possibilities further usually increases exponentially. Thus science can only afford the luxury of determining the exact molecular structure of a new unknown compound for special cases. Knowledge of even a few of the gross structural features often is enough to tell the scientist that the compound is not germane to his investigation--for example, that it is probably non-toxic, should not show the desired pharmacological activity, or is not a logical metabolite of the drug used.

Outside Use of STIRS and PBM. STIRS has been available free to outside users since January 1974 through a phone-line link to our laboratory PDP-11/45, and has been used at a rate of ~100 unknown spectra per month for a substantial part of this time. This indicated that STIRS was uniquely meeting a real scientific need, but also showed that a very substantial proportion of the unknowns should have been examined first by a retrieval system. This led to implementing both PBM and STIRS on the Cornell IBM 370/168 to make them available internationally over the TYMNET computer network system. This system has the further advantage that it employs a data base that currently has approximately 40,000 mass spectra of 30,000 different compounds (15) which, to our knowledge, is larger than any other available collection.

The Cornell PBM/STIRS system became operational on TYMNET in May 1975 and the initial response has been encouraging. The programs appear to be used in a complementary fashion; if PBM cannot identify the unknown mass spectrum as a compound already in the reference file at a satisfactory confidence level, STIRS usually elucidates at least partial structural information concerning the unknown. Not surprisingly, especially high enthusiasm has been expressed by those laboratories with relatively little experience in mass spectral interpretation. Although highly automated gas chromatograph/mass spectrometer/computer systems are common in many service laboratories, such as for drug identification, forensic analysis, and clinical assays, the spectral retrieval systems available with these are relatively primitive, usually based on a limited number of peaks, and utilizing only small specialized data bases. More experienced mass spectrometry laboratories seem to be using the Cornell system mainly for "difficult" spectra, finding PBM valuable because of our more comprehensive data base, and STIRS of special use for those types of compounds with which the laboratory is not particularly familiar.

The Heller/NIH "Conversational Mass Spectral Search System" (19) has been available for phone-line use since 1971, and since 1973 over the GE international computer network. The present list of over 200 users attests to the need as well as the usefulness of this system. At the time of this writing, the system utilized only approximately 13,000 reference spectra, but enlargement of the data base to 35,000 spectra is planned for the immediate future. This interactive system can be used for both retrieval and interpretive purposes, interrogating the reference file concerning the presence of particular peaks on an individual basis. Although this system is not suitable for automatic matching of complete unknown mass spectra, such as is PBM, a detailed comparison of the advantages and disadvantages of this conversational system versus the PBM/STIRS system has not been made.

Applicability of a Computer Networking System to Analysis of Unknown Spectra. Our experience to date does give us some indication of the types of problems which are best examined by such computer systems. It seems logical that the best way to implement a retrieval system is as part of the computer data acquisition and reduction system of the mass spectrometer. Here the practical limitations of computer speed, power, and reference data storage capacity will prevent such a system from being used in particular applications. However the application of PBM in a GC/MS system controlled by a dedicated microcomputer (12) shows that this is certainly the method of choice under the proper circumstances. Further, if such a system will produce answers quickly and accurately in a substantial proportion of cases, the remainder can then obviously be examined in the more sophisticated PBM/STIRS system available on the network.

It does appear that the ready accessibility to most laboratories which will on occasion require such a service is an important key to the success of such a centralized system for mass spectral retrieval and interpretation. It is very difficult for most laboratories to maintain a growing mass spectral data base of high accuracy. Further, systems such as PBM and STIRS are undergoing rapid research improvements and obviously there must be a large time lag in incorporation of such changes unless there is a close connection with the research laboratory involved. The proper method of funding such a centralized resource has not been resolved; at present the financial support of the Cornell PBM/ STIRS system is entirely based on computer use charges, with further research and development of the systems supported through Federal grants. A number of users have expressed strong opinions that the whole operation should be funded by the Federal Government, with at least academic and governmental users paying at most a small service charge for the networking operation.

Acknowledgment. The authors are deeply indebted to the National Institutes of Health, the National Science Foundation, and the Environmental Protection Agency for generous financial support of the research programs on PBM and STIRS. The PBM system was developed in cooperation with Dr. R. H. Hertel and R. D. Villwock of the Universal Monitor Corporation, Pasadena, California. Implementation of PBM and STIRS on the network system would have not have been possible without the close cooperation of J. W. Rudan, J. Aikin, R. Cogger, and B. A. Meyer of the Office of Computer Services, Cornell University.

Literature Cited

1. Ridley, R. G., in Waller, G. R., Ed., "Biochemical Applications of Mass Spectrometry," p 177, John Wiley, New York City, 1972.
2. Fennessey, P. V., in Milne, G. W. A., Ed., "Mass Spectrometry: Techniques and Applications," p 77, John Wiley, New York City, 1971.
3. Fenselau, C., Appl. Spectr., (1974), 28, 305.
4. Arpino, P. J., Dawkins, B. G., and McLafferty, F. W., J. Chrom. Sci., (1974), 12, 574.
5. Pesyna, G. M., and McLafferty, F. W., "Determination of Organic Structures by Physical Methods," Volume 6, Nachod, F. C., Zuckerman, J. J., and Randall, E. W., Academic Press, New York City, 1975.
6. Crawford, L. R., and Morrison, J. D., Anal. Chem., (1968), 40, 1464.
7. Jurs, P. C., Kowalski, B. R., and Isenhour, T. L., Anal. Chem., (1969), 41, 21.
8. Isenhour, T. L., and Jurs, P. C., Anal. Chem., (1971), 43, 21A.
9. Buchs, A., Delfino, A. B., Duffield, A. M., Djerassi, C., Buchanan, B. G., Feigenbaum, E. A., and Lederberg, J., Helv. Chim. Acta, (1970), 53, 1394.
10. Smith, D. H., Buchanan, B. G., Engelmore, R. S., Duffield, A. M., Yeo, A., Feigenbaum, E. A., Lederberg, J., and Djerassi, C., J. Am. Chem. Soc., (1972), 94, 5962.
11. Kwok, K.-S., Venkataraghavan, R., and McLafferty, F. W., J. Am. Chem. Soc., (1973), 95, 4185.
12. McLafferty, F. W., Hertel, R. H., and Villwock, R. D., Org. Mass Spectrom., (1974), 9, 690.
13. Pesyna, G. M., McLafferty, F. W., Venkataraghavan, R., and Dayringer, H. E., Anal. Chem., submitted.
14. Stenhagen, E., Abrahamsson, S., and McLafferty, F. W., "Registry of Mass Spectral Data," Wiley-Interscience, New York City, 1974.
15. Mass Spectral Data Collection, Cornell University and the Environmental Protection Agency.
16. Salton, G., "Automatic Information, Organization, and Retrieval," McGraw-Hill, New York City, 1968.
17. Pesyna, G. M., McLafferty, F. W., and Venkataraghavan, R., Anal. Chem., June, (1975).

18. Dayringer, H. E., Pesyna, G. M., Venkataraghavan, R., and McLafferty, F. W., Anal. Chem., submitted.
19. Heller, S. R., Koniver, D. A., Fales, H. M., and Milne, G. W. A., Anal. Chem., (1974), 46, 947.

13

Networking and a Collaborative Research Community: a Case Study Using the DENDRAL Programs

RAYMOND E. CARHART, SUZANNE M. JOHNSON, DENNIS H. SMITH, BRUCE G. BUCHANAN, R. GEOFFREY DROMEY, and JOSHUA LEDERBERG

Departments of Computer Science, Genetics, and Chemistry, Stanford University, Stanford, Calif. 94305

Computer Science is one of the newest, but also one of the least "cumulative" of the sciences. Gordon (1) has recently pointed out the upsetting disparity between the number of potentially sharable programs in existence and the number which are easily accessible to a given researcher. Although some mechanisms exist for the systematic exchange of program resources, for example the World List of Crystallographic Computer Programs (2), a great deal of programming effort is duplicated among different research groups with common interests. The reasons for this are understandable: these groups are separated by geography, by incompatibilities in computer facilities and by a lack of a means to keep abreast of a rapidly changing field.

The emergence of more economical technologies for data communications provides, in principle, a method for lowering these geographical and operational barriers; for creating, through computer networking, remote sites at which functionally specialized capabilities are concentrated. The SUMEX-AIM (Stanford University Medical EXperimental computer - Artificial Intelligence in Medicine) project is an experiment in reducing this principle to practice, in the specific area of artificial intelligence research applied to health sciences.

The SUMEX-AIM computer facility (3) is a National Shared Computing Resource being developed and operated by Stanford University, in partnership with and with financial support from the Biotechnology Resources Branch of the Division of Research Resources, National Institutes of Health. It is national in scope in that a major portion of its computing capacity is being made available to authorized research groups throughout the country by means of communications networks.

Aside from demonstrating, on managerial, administrative and technical levels, that such a national computing resource is a viable concept, the primary objective of SUMEX-AIM is the building of a collaborative research community. The aim is to encourage individual participants not only to investigate applications of artificial intelligence in health science, but also to share their programs and discuss their ideas with other researchers. This places a responsibility upon SUMEX-AIM to develop effective means of communication among community members and among the programs they write. It also places responsibility upon those members to design and document programs that readily can be used and understood by others.

Another aspect of the SUMEX facility is providing service to individuals whose interest is in using, rather then developing, the available computer programs. Although this is not a primary consideration, it is an essential part of the growth of these programs. Most of the SUMEX-AIM projects have formed, or are forming, their own user communities which provide valuable "real world" experience. Figure 1 depicts the typical interaction of such a project with its user community, and with other projects. The participation by users in program development is not just restricted to suggestions, but can also include software created by computer-oriented users to satisfy special needs. In some projects, methods are being considered to further promote this kind of participation.

The purposes of this paper are threefold: first, to indicate the range of research projects currently active at SUMEX; second, to describe in detail one of these projects, DENDRAL, which is of particular interest to chemists; and third, to discuss some problems and possible solutions related to networking and community-building.

I. RESEARCH ACTIVITIES AT SUMEX-AIM

The community of participants in SUMEX-AIM can be divided geographically into local (i.e., Stanford-based) projects and remote projects, and below is given a brief description of the major representatives of each. Communication with the remote projects is accomplished through one or both of the communications networks shown in Figure 2. In most cases, connection with SUMEX-AIM from these remote sites involves only a local telephone call to the nearest network "node".

The SUMEX-AIM system is itself undergoing constant improvement which deserves to be called research, and thus a third section is included here to represent system development.

Remote Projects

The Rutgers Project. Originating from Rutgers University are several research efforts designed to introduce advanced methods in computer science - particularly in artificial intelligence and interactive data base systems - into specific areas of biomedical research. One such effort involves the development of computer-based consultation systems for diseases of the eye, specifically the establishment of a national network of collaborators for diagnosis and recommendations for treatment of glaucoma by computer.

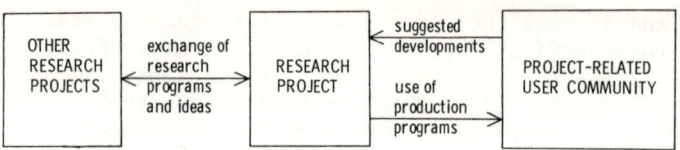

Figure 1. Interactions in the SUMEX-AIM Community

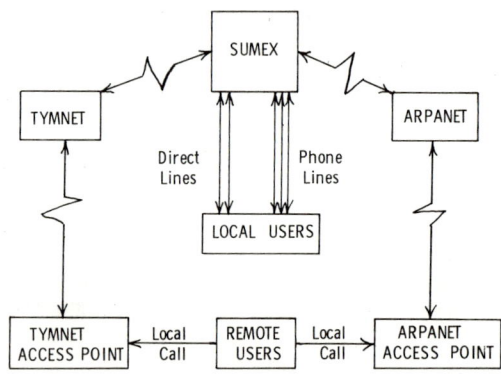

Figure 2. Access to SUMEX-AIM

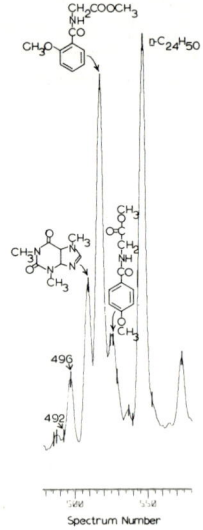

Figure 3. Total ion current vs. spectrum number in a GC/LRMS run

Another project concerns the BELIEVER program, which represents a theory of how people arrive at an interpretation of the social actions of others. SUMEX-AIM provides an excellent medium for collaboration in the development and testing of this theory. The Rutgers project includes, in addition, several fundamental studies in artificial intelligence and system design, which provide much of the support needed for the development of such complex systems.

The DIALOG Project. The DIAgnostic LOGic project, based at the University of Pittsburgh, is a large scale, computerized medical diagnostic system that makes use of the methods and structures of artificial intelligence. Unlike most other computer diagnostic programs, which are oriented to differential diagnosis in a rather limited area, the DIALOG system has been designed to deal with the general problem of diagnosis in internal medicine and currently accesses a medical data base which encompasses approximately fifty percent of the major diseases in internal medicine.

The MISL Project. The Medical Information Systems Laboratory at the University of Illinois (Chicago Circle campus) has been established to explore inferential relationships between analytic data and the natural history of selected eye diseases, both in treated and untreated forms. This project will utilize the SUMEX-AIM resource to build a data base which could then be used as a test bed for the development of clinical decision support algorithms.

Distributed Data-Base System for Chronic Diseases. This project, based at the University of Hawaii, seeks to use the SUMEX-AIM facility to establish a resource sharing project for the development of computer systems for consultation and research, and to make these systems available to clinical facilities from a set of distributed data bases. The radio and satellite links which compose the communication network known as the ALOHANET, in conjunction with the ARPANET, will make these programs available to other Hawaiian islands and to remote areas of the Pacific basin. This project could well have a significantly beneficial effect on the quality of health care delivery in these locations.

Modelling of Higher Mental Functions. A project at the University of California at Los Angeles is using the SUMEX-AIM facility to construct, test, and validate an improved version of the computer simulation of paranoid processes which has been developed. These simulations have clinical implications for the understanding, treatment, and prevention of paranoid disorders. The current interactive version (PARRY) has been running on SUMEX-AIM and has provided a basis for improvement of the future version's language capability.

Local Projects

The Protein Crystallography Project. The Protein Crystallography project involves scientists at two different universities (Stanford and the University of California at San Diego), pooling their respective talents in protein crystallography and computer science, and using the SUMEX-AIM facility as the central repository for programs, data and other information of common interest. The general objective of the project is to apply problem solving techniques, which have emerged from artificial intelligence research, to the well known "phase problem" of x-ray crystallography, in order to determine the three-dimensional structures of proteins. The work is intended to be of practical as well as theoretical value to both computer science (particularly artificial intelligence research) and protein crystallography.

The MYCIN Project. MYCIN is an evolving program that has been developed to assist physician nonspecialists with the selection of therapy for patients with bacterial infections. The project has involved both physicians, with expertise in the clinical pharmacology of bacterial infections, and computer scientists, with interests in artificial intelligence and medical computing. The MYCIN program attempts to model the decision processes of the medical experts. It consists of three closely integrated components: the Consultation System asks questions, makes conclusions, and gives advice; the Explanation System answers questions from the user to justify the program's advice and explain its methods; and the Rule-Acquisition System permits the user to teach the system new decision rules, or to alter pre-existing rules that are judged to be inadequate or incorrect.

The DENDRAL Project. This project, being of particular chemical interest, is described in detail in Section II. Through the SUMEX-AIM facility DENDRAL has gained a growing community of production-level users whose experience with the programs is a valuable guide to further development. Although technically users, some members of this community might better be described as collaborators because they have provided SUMEX-AIM with various special-purpose programs which are of interest to other chemists and which extend the usefulness of the DENDRAL programs.

SUMEX-AIM System Development

Current research activities at SUMEX-AIM are developing along several lines. On a system development level there are ongoing projects designed to make the system more user oriented. Currently, the system can be expected to provide help to the user who is confused about what is expected in response to a certain prompt. A "?" typed by the user, will, in most cases, provide a list of possible responses from which to choose. Also available in response to typing "HELP" to the monitor is a general help description containing pointers to files likely to be of interest to a new user.

In an effort to facilitate communication between collaborators, a program called CONFER has been developed to provide an orderly method for multiple participant teletype "conference calls". Basically, the program acts as a character processor for all the terminals linked in the conference, accepting input from only one at a time, and passing it out to the remaining terminals. In this way, the conference, in effect, has a "moderator" terminal, thus allowing for a more orderly transfer of ideas and information.

SUMEX-AIM is also aware of the necessity of making its facilities available for trial use by potential users and collaborators. To this end, a GUEST mechanism has been established for persons who wish to have brief, trial access to certain programs they feel may be of value to them, and about which they would like to obtain more knowledge. This provides a convenient mechanism whereby persons, who have been given an appropriate phone number and LOGIN procedure, can dial up SUMEX-AIM and receive actual experience using a program they may only have heard about.

Another area of system development currently being explored at SUMEX-AIM is that of creating a comprehensive "bulletin board" facility where users can file "bulletins", that is, messages of interest to the SUMEX-AIM community. The facility will also alert users to new bulletins which are likely to be of interest to them, as determined by individual user-interest profile.

II. DENDRAL - CHEMICAL APPLICATIONS OF INTERACTIVE COMPUTING IN A NETWORK ENVIRONMENT

The major research interest of the DENDRAL Project at Stanford University is application of artificial intelligence techniques for chemical inference, focusing in particular on molecular structure elucidation. Portions of our research are in the area of combined gas chromatography/high resolution mass spectrometry and include instrumentation and data acquisition hardware and software development. This area is beyond the scope of this report; we focus instead on the concurrent development of programs to assist chemists in various phases of structure elucidation beyond the point of initial data collection. SUMEX-AIM provides the computer support for development and application of these programs.

Another aspect of our research is our commitment to share developments among a wider community. We feel that several of our programs are advanced enough to be useful to chemists engaged in related work in mass spectrometry and structure elucidation in general. These programs are written primarily in the programming language INTERLISP, and thus are not easily exportable (exceptions are indicated subsequently). SUMEX-AIM provides a mechanism for

allowing others access to the programs without the requirement for
any special programming or computer system expertise. The availability of the SUMEX facility over nationwide networks allows
remote users to access the programs, in many instances via a local
telephone call.

Much of the following discussion is preliminary because our
programs have only recently been released for outside use. Some
announcement of their availability has been made, and other
announcements will occur in the near future, through talks, publi-
cations in press, demonstrations and informal discussions.
Although most of our experience has been with local users, they
have been good models of remote users in that their previous expo-
sure to the actual programs and computer systems is minimal. Their
experience has been extremely useful in helping us to smooth out
clumsy interactions with programs and to locate and fix program
bugs. Such polishing is important for programs which may be utilized by users from widely differing backgrounds with respect to
computers, networks and time sharing systems. We are in the processes of building a community of remote users. We actively
encourage such use for two reasons: 1) we feel the programs are
capable of assisting others in solving certain molecular structure
problems, and 2) such experience with outside users will be a
tremendous assistance in increasing the power of our programs as
the programs are forced to confront new real-world problems.

The remainder of this section outlines the programs which are
available via SUMEX, the utilization of these programs in helping
to solve structure elucidation problems and the limitations we see
to their use. We discuss current applications of the programs to
our research and the research of other users to illustrate better
the variety of potential applications and to stimulate an inter-
change of ideas. Where appropriate, we point out current difficulties with the use both of our programs and of SUMEX. New
applications and wider use will certainly change the nature of
these problems; we strive to solve current problems, but new ones
will always arise to take their place.

DENDRAL Programs

We have several programs which we employ in dealing with
various aspects of problems involving unknown structures. Some of
these programs are exportable, while the remainder are available
at SUMEX. The availability of each program is discussed below.

Our initial emphasis in studying applications of artificial
intelligence for chemical inference was in the area of mass spectrometry (4-6). This emphasis remains because many of our problems require mass spectrometry as the analytical tool of choice in
providing structural information on small quantities of sample.
More recently, we have been developing a program (CONGEN, below)
directed at more general aspects of structure elucidation. This
has extended the scope of problems for which we can provide

computer assistance.

We will begin, however, with discussion of the mass spectrometry programs. The examples used in the discussion are characteristic of our current research problems, although we have focused on relatively simple problems to keep the presentation brief. We trace, in what might be chronological terms, the application of the programs to various phases of a structure problem. In this way we hope to illustrate the place of each program in the analysis. We begin by discussing preprocessing of mass spectral data (CLEANUP and MOLION). Subsequent analysis of such data in terms of structure is then covered (PLANNER). The use of CONGEN is discussed for problems which cannot be handled by the previous programs. Finally, we discuss efforts to discover, with the use of the computer, systematics in the behavior of known substances in the mass spectrometer as a means of extending the knowledge of the system for applications in new areas (INTSUM and RULEGEN).

Programs for Molecular Structure Problems

The first three programs, CLEANUP, the library-search program and MOLION are in a sense utility programs, but all three play a critical role in processing mass spectral data. Subsequent applications of programs (e.g., PLANNER) for more detailed spectral analysis in terms of structure depend on the successful treatment of the data by CLEANUP and MOLION, while the library search program filters out common spectra which need not undergo a full analysis. The examples used are drawn from our collaboration with persons in the Genetics Research Center at Stanford Hospital. The experimental data which are collected are the results of combined gas chromatographic/low resolution mass spectral (GC/LRMS) analysis of organic components (chemically fractionated and derivatized where necessary) of body fluids, e.g. blood, urine. A typical experiment consists of 500-600 individual mass spectra for each fraction, taken sequentially over time as the various components, largely separated from one another, elute from the gas chromatograph and pass into the mass spectrometer. Each mass spectrum consists of the mass analyzed fragment ions of the component(s) in the mass spectrometer at the time the spectrum was taken. Such spectra are related, indirectly, to the molecular structure of the component(s).

CLEANUP(7). The individual mass spectra obtained from fractionated GC/LRMS analysis are quite often poor representations of corresponding spectra taken from pure compounds. They can be contaminated by the presence of additional peaks and/or distortions of the intensities of existing peaks in the spectrum. Fragment ions from either the liquid phase of the GC column or from components incompletely separated by the gas chromatograph are responsible for the contamination. We have developed a program, referred to here as CLEANUP, which examines all mass spectra in a

GC/LRMS run, selects those spectra which contain ions other than
background impurities, and removes contributions from background
and overlapping components. A spectrum results which compares
favorably with the spectrum of a pure component. Biller and
Biemann (8) have developed a similar but less powerful program.

For example, the CLEANUP program detected components at
points marked with a vertical bar in the (partial)plot of total
ion current vs. scan number (time), Figure 3. Note that overlapping components were detected under the envelopes of the GC peaks
in the region of scans 485-488, 525-529 and 539-552. We focus our
attention on the spectrum recorded at scan 492. The raw data,
prior to cleanup, are presented in Figure 4 (top). The spectrum
resulting from CLEANUP is presented in Figure 4 (bottom). Note
that the large ions (e.g., m/e 207, 221 and 315) from background
impurities are removed, and that the intensity ratios of peaks at
lower masses (e.g., 51 and 77) have been adjusted to reflect their
true intensities in the spectrum.

The CLEANUP program is capable of detection of quite
low-level components in complex mixtures as indicated by some of
the areas of the total ion current plot (Figure 3) where components were detected. It is completely general because nothing in
the program code is sensitive to the types of compounds analyzed
or the characteristics of possible impurities associated with the
compounds or from the GC column. Its major limitation is that
mass spectra must be taken repetitively during the course of a
GC/MS run. Its performance is enhanced when such spectra are
measured closely in time.

The program is offered via SUMEX as an adjunct to use of our
other programs; it is not offered as a routine service. Because
the program is written in FORTRAN, we routinely use it on our data
acquisition computer system so as not to burden SUMEX with tasks
better done elsewhere. Similarly, we would assist other frequent
users to mount the program on their own systems.

Library Search. With a set of "clean" mass spectra
available, the next problem is identification of the various
components. Over the course of several years, libraries of mass
spectral data have been assembled (9). These libraries can be
very useful in weeding out from a group of spectra those which
represent known compounds (10). Clearly, one should spend time on
solving the structures of unknown compounds, not on rediscovering
old ones. The CLEANUP program provides mass spectra which are of
sufficient quality to expect that known compounds would be identified easily from such libraries.

The spectra detected by CLEANUP in the region of scans 480 to
580 (Figure 3) were matched against the library of biomedically
relevant spectra compiled by S. Markey (National Institutes of
Health) and our extensions to that library (we wish to thank S.
Grotch, Jet Propulsion Laboratory, Pasadena, Ca. for providing
some of the library matchings). Excellent matches with the

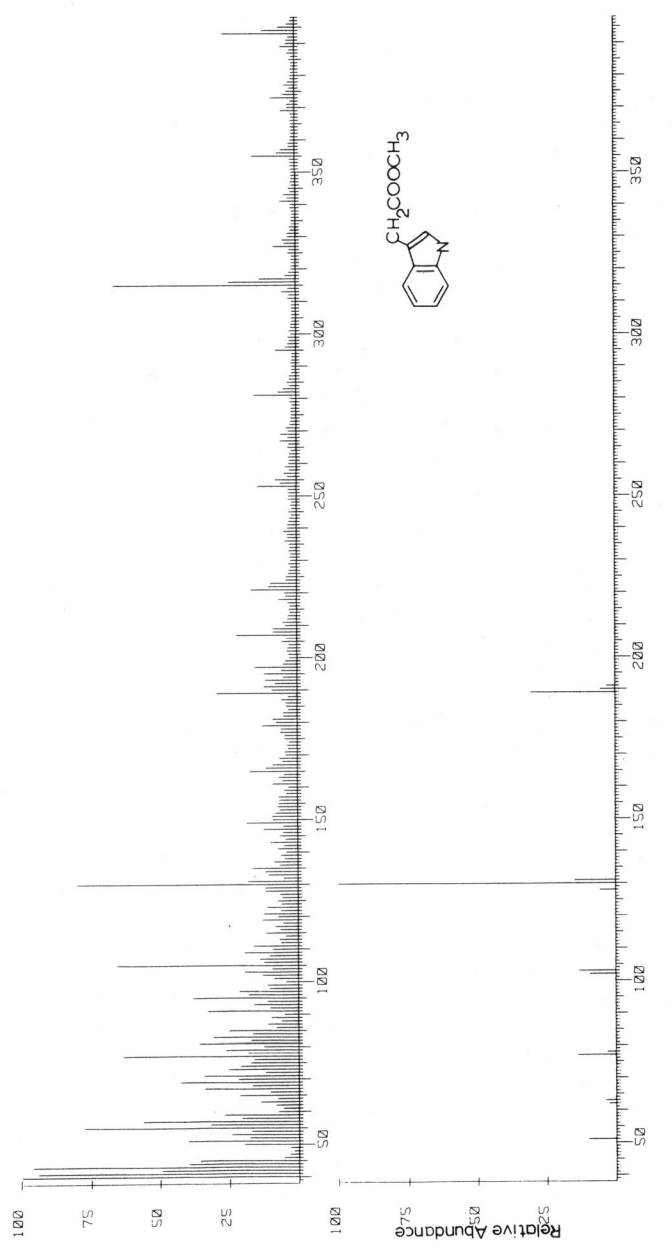

Figure 4. Spectrum number 492 from the GC/LRMS trace shown in Figure 3: (top) raw data; (bottom) spectrum output by CLEANUP

library were obtained for scans 492, 496, 509, 519, 529 and 548. The components are indole acetic acid methyl ester, di-n-butlyphthalate, caffeine, salicyluric acid methyl ester, methoxyhippuric acid methyl ester and n-C24 hydrocarbon respectively (structures given in Figures 3 and 4). Spectra scans at 485, 487, 525, 530, 536, 539, 554 and 576 did not match well with any spectrum in the library and thus must be examined further for structuring information. The necessity for preprocessing the data using CLEANUP prior to library matching is illustrated from indole acetic acid methyl ester (scan 492). The "clean" spectrum (Figure 4, bottom) was matched to the library spectrum of this compound much better than to that of any other compound. The raw spectrum (Figure 4, top), when compared to the library resulted in eleven other compounds which matched more closely that the correct one.

This brief example illustrates the obvious value and limitations of library searching. The most interesting compounds for subsequent analysis are those which are unknown. The fractions of urine extracts are replete with unidentified compounds because of the inadequacy of current library compilations. As new compounds are identified they are, of course, added to the library so that future analyses need not reinvestigate the same material.

We currently perform library searching on our data acquisition and reduction computer systems. We can, if necessary, offer limited library search facilities via SUMEX. However, because commercial facilities are available (e.g., over the GE network), routine library search service is not available on SUMEX.

MOLION(11). At this stage we are left with a collection of mass spectra of unknown compounds. The library search results may have provided some clues as to the type of compound present, e.g., compound class. Structure elucidation now begins in earnest. The key elements in problems of structure elucidation are the molecular weight and empirical formula of a compound. Without these essential data, the structural possibilities are usually too immense to proceed further. Mass spectrometry is frequently used to determine molecular weights and formulae, but there is no guarantee that the mass spectrum of a compound displays an ion corresponding to the intact molecule. For example, many of the derivatives of the amino acid fractions of urine display no molecular ions. When we are given only the mass spectrum (and for GC/MS analysis a mass spectrum may be all that is available) we must somehow predict likely molecular ion candidates. The program MOLION performs this task. Given a mass spectrum, it predicts and ranks likely molecular ion candidates independent of the presence or absence of an ion in the spectrum corresponding to the intact molecule. The published manuscript (11) provides many examples of the performance of the program.

The mass spectrum of an example, unknown X, (which we will pursue in more detail below) is given in Figure 5. The results obtained from MOLION are summarized in Table I. The observed ion at m/e 263 is ranked as the most likely candidate.

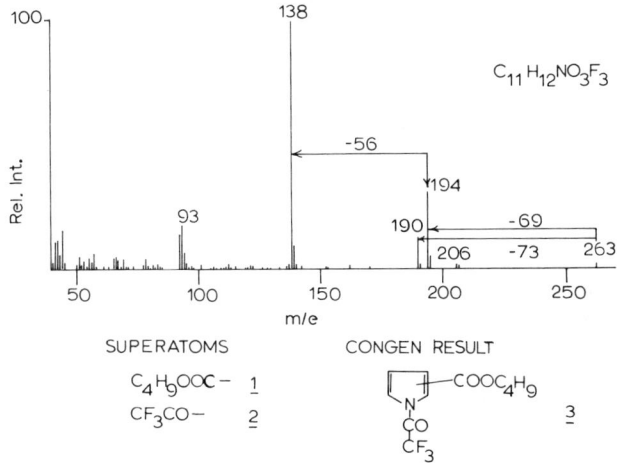

Figure 5. *Low-resolution mass spectrum of unknown X. The indicated superatoms were deduced from the spectrum and the chemical history of the sample. Based on these and other constraints, CONGEN obtains the indicated result.*

Table I. Results of Molecular Ion Determination for the Unknown Compound, X, whose Mass Spectrum is Presented in Figure 5.

CANDIDATE	RANKING INDEX
263.0	100
307.0	41
299.0	38
295.0	34
281.0	25

The MOLION program is written to operate on either low or high resolution mass spectra. The program has certain limitations which have been summarized in detail previously (11).

MOLION is available on SUMEX. A FORTRAN version, initially for low resolution mass spectra, is being written so that the program can be run on smaller computers and exported to others. However, it will continue to be available via SUMEX so that others can access it easily. MOLION is contained within PLANNER as one of the available methods for detecting candidate molecular ions.

PLANNER(12). The PLANNER program is designed to analyze the mass spectrum of a compound or of a mixture of related compounds. Because there is no *ab initio* way of relating a mass spectrum of a complex organic molecule to the structure of that molecule, PLANNER requires fragmentation rules for the class of compounds to which the unknown belongs. This is its major limitation. For our example the class was unknown, forcing us to resort to other means of assistance.

Applications and limitations of PLANNER have been discussed extensively (12,13). The program is very powerful in instances where mass spectrometry rules are strong (*i.e.*, general, with few exceptions). In instances where rules are weak or nonexistent, additional work on known structures and spectra may yield useful rules to make PLANNER applicable (see INTSUM and RULEGEN, below). One important feature of PLANNER is its ability to analyze the spectra of mixtures in a systematic and thorough way. Thus, it can be applied to spectra obtained as mixtures when GC/MS data are unavailable or impossible to obtain. PLANNER is available in an interactive version over SUMEX, requiring three kinds of information as input: the high or low resolution mass spectrum, the characteristic skeletal structure for molecules in the specific compound class, and the fragmentation rules for the class. Additional knowledge about the unknown can be used by the program to constrain the structural possibilities.

CONGEN(14,15). Structure problems are usually not solved with mass spectrometry alone. Even when sample size is too limited for obtaining other spectroscopic data, knowledge of chemical isolation and results of derivatization procedures frequently act as powerful constraints on structural possibilities. Larger amounts of sample permit determination of other spectroscopic data. Taken together, this information allows determination of structural features (substructures) of the molecule and constraints on the plausibility of ways in which the substructures may be assembled. The CONGEN program is capable of providing assistance in solution of such problems.

CONGEN performs the task of construction, or generation, of structural isomers under constraints. The program accepts as input known structural fragments of the molecule ("superatoms") and any remaining atoms (C,N,O,P,...), together with constraints on how they may be assembled. It is based on the exhaustive structure generator (16,17) and extensions (18) which permit a stepwise assembly of structures.

In an interactive session with the program, a user supplies structural information determined by his own analysis of the data (perhaps with the help of the above programs), together with whatever other constraints are available concerning desired and undesired structural features, ring sizes and so forth. The program builds structures in a series of steps, during which a user can interact further with the procedure, for example, to add new constraints. Although very much a developing program, its ability to accept user-inferred constraints from many data sources makes CONGEN a general tool for structure elucidation which we are making available via SUMEX-AIM in its current form.

For the unknown X, the observed fragment ions from the molecular ion (M) at m/e 263 (Figure 5) suggest several structural features when coupled with the knowledge of the chemical derivatization procedures used on this fraction of the urine extract. The ion at m/e 194 represents loss of 69 amu, probably CF3, from fragmentation of a trifluoroacetyl derivative of an amine. This suggests the partial structure 2, Figure 5. The ions at m/e 190 (M-74 amu) and m/e 162 (M-101 amu) suggest the characteristic fragmentation of an n-butyl ester resulting from the second derivatization procedure, formation of the n-butyl esters of free carboxylic acid functions. This suggests the partial structure 1, Figure 5. Taken together, all the above information implies (if no other elements are present) that the empirical formula contains an odd number of nitrogen atoms, at least three oxygen atoms, three fluorine atoms and at least seven carbon atoms. Interestingly, there is only one plausible empirical formula under these constraints, C11H12NO3F3.

Structural fragments ("superatoms") 1 and 2 were supplied to CONGEN, together with the remaining four carbon atoms and three degrees of unsaturation (that is, rings plus multiple bonds). With no additional constraints, 155 structures result. The inclusion of other plausible constraints (e.g., no allenes, acetylenes, cyclopropenes, cyclobutenes) reduces the number of structural candidates to just the two isomeric forms of 3, Figure 5.

This problem represents a simple example of a large class of such problems. Although a chemist could probably reach the same conclusions quickly in this case, in the general case, piecing together potential solutions is not a trivial task.

Although still a developing program, CONGEN is capable of considerable assistance in a wide variety of structure problems. Some areas of current application are summarized in the subsequent section. It is already proving its value in structure elucidation problems by suggesting solutions with a guarantee that no plausible alternatives have been overlooked.

The program has a great deal of flexibility. Many of the types of constraints normally brought to bear on structure elucidation problems can be expressed. However, some types of constraints cannot be easily expressed (e.g., disjunctions of features and stereo-constraints). Recent work by our group and Wipke's (19) will make it possible to add considerations of stereoisomerism relatively easily (a good example of collaboration via SUMEX). We are depending on a broad user community to help us guide further development of CONGEN.

Programs for Knowledge Acquisition

INTSUM(20) and RULEGEN(21). When the mass spectrometry rules for a given class of compounds are not known, the INTSUM and RULEGEN programs can help a chemist formulate those rules. Essentially, these programs categorize the plausible fragmentations for a class of compounds by looking at the mass spectra of several molecules in the class. All molecules are assumed to belong to one class whose skeletal structure must be specified. Also, the mass spectra and the structures of all the molecules must be given to the program.

INTSUM collects evidence for all possible fragmentations (within user-specified constraints) and summarizes the results. For example, a user may be interested in all fragmentations involving one or two bonds, but not three; aromatic rings may be known to be unfragmented; and the user may be interested only in fragmentations resulting in an ion containing a heteroatom. Under these constraints, the program correlates all peaks in the mass spectra with all possible fragmentations. The summary of results shows the molecules whose spectra display evidence for each particular fragmentation, along with the total (and average) ion current associated with the fragmentation.

The RULEGEN program attempts to explain the regularities found by INTSUM in terms of the underlying structural features around the bonds in question that seem to "direct" the fragmentations. For example, INTSUM will notice significant fragmentation of the two different bonds alpha to the carbonyl group in aliphatic ketones. It is left to RULEGEN to discover that these are both instances of the same fundamental alpha-cleavage process that can be predicted any time a bond is alpha to a carbonyl group.

These programs are part of the so-called Meta-DENDRAL effort, whose general goal is to understand rule formation activities. Both INTSUM and RULEGEN are available as interactive programs on SUMEX, the former being much more highly developed that the latter. Although these programs can be very useful to chemists interested in finding new mass spectrometry rules, they require having the collection of mass spectra and molecular structure descriptions available in one computer file. Because of this, they have been used mostly by chemists at Stanford.

Applications and Resource Sharing

The DENDRAL programs are being developed to serve a broad community of chemists with structure elucidation problems. Our experience is admittedly limited. In this section we discuss some of the applications, both local and from remote sites, where these programs have proven useful.

CLEANUP and MOLION. These programs are in routine use as part of the Genetics Research Center's GC/LRMS efforts. In addition, MOLION has been incorporated as part of PLANNER. Their generality has proven very useful in applications to a variety of GC/MS problems involving structural studies of urinary metabolites.

PLANNER. The planning program has been used to infer plausible placement of substituents around a skeletal structure for numerous test problems in which the class of the sample was known and the fragmentation rules for the class were known. Those tests have resulted in a program that we believe is general. We have applied this program to unknown mixtures of estrogenic steriods (13). We are preparing to use PLANNER for screening mass spectra of marine sterols to identify quickly those spectra of known compounds and to suggest structures for spectra of new compounds.

CONGEN. CONGEN is being used locally and from remote sites in a wide variety of applications. We have used it for construction of ring systems under constraints (22) and for generation of structures of chlorocarbons (23). We have investigated several monoterpenoid and sesquiterpenoid structure problems to suggest solutions and to ensure that all alternatives had been considered. We are currently investigating the scope of terpenoid isomerism. Two problems relating to unknown photochemical reaction products have been analyzed and results used to suggest further experiments. In most cases we do not know the precise problems under study by remote users, only that they are using the program.

CONGEN will perhaps be the most widely used (by remote users) program of those mentioned above as accessible through SUMEX. This is primarily a result of the wider scope of problems which might benefit from use of the program. However, the need for remote users to have their mass spectral data available at SUMEX for analysis present a significant energy barrier to use of the programs which require these data.

INTSUM and RULEGEN. INTSUM is essentially a production program now, and is being used as such in a variety of applications involving correlations of molecular structures with their respective spectra. Recent or current applications include analysis of the mass spectra of progesterones and related steroids, androstanes, macrolide antibiotics, insect juvenile hormones and phytoecdysones. These studies serve to develop fragmentation rules which, if of sufficient generality, can in turn be used in PLANNER in the study of unknown compounds.

III. PROBLEMS RELATED TO NETWORKING

During this first year of operation, the SUMEX-AIM facility has encountered a variety of problems arising from its network availability. In most cases, there has been no clear precedent

for the handling of these situations, in fact, many problem-areas
still reflect the influences of a yet-developing policy. The hope
is that this presentation and discussion of problems and their
solutions may give foresight to others who contemplate networking
or network use. The problems to be discussed can be loosely
associated into three classes; those related to the management of
the facility, those pertaining to research activities on the system, and those involving psychological barriers to network use.

Managerial Problems

"Gatekeeping". The most general problem faced by the
organizers of the SUMEX-AIM facility is the question of "gatekeeping." In order to insure a high quality of pertinent research,
some kind of refereeing system is needed to assess the value of
proposed new projects. The organizers of the facility would seem
to be the best source of such judgements; yet, because we are both
organizers and members of the SUMEX community, there is a danger
that our decisions would unfairly favor local priorities. In
order to establish credibility in SUMEX-AIM as a truly national
resource, a management system has been instituted that allocates a
defined fraction (initially 50%) of the SUMEX resource to external
users, under the jurisdiction of an independent national committee
(the AIM advisory group). The remaining 50%, allocated for local
use, contains a portion for flexible experiments outside of local
projects, but on our own responsibility.

Choice of computer and operating system. A second management
level problem is the choice of a computer and operating system
which optimize the usefulness of the facility for a majority of
users, and which encourage intercommunication between remote collaborators. Because SUMEX-AIM is intended to be used primarily
for applications of artificial intelligence, and because interactive LISP (24) is a primary language in this type of work, the
choice of TENEX (25) as an operating system was dictated somewhat
by necessity. TENEX incorporates multiple address spaces, thereby
allowing multiple "fork" structure and paging, a design which is
necessary to create the large memory virtual machine required by
INTERLISP.
    The PDP-10 is a popular machine for interactive computing of
all sorts in university research environments, and thus an added
benefit of this choice was expected - the possibility of easily
transferring to SUMEX programs developed at other sites. Many of
these programs were written not under TENEX but under the 10/50
monitor supplied by the manufacturer. Because a large and useful
program library was already available under the 10/50 monitor, one
of the design criteria of TENEX was compatibility with such programs; when a 10/50 program is run under TENEX, a special "compatibility package" of routines is invoked to translate 10/50 monitor
calls into equivalent TENEX monitor calls. Although the concept

is sound, we have found that in practice very few programs
written for the 10/50 monitor are able to run under TENEX without
extensive modification. Other problems with TENEX include weak-
nesses in the support of peripheral devices and the lack of a
default line-editor. The latter has caused a proliferation of
editing programs, and some confusion has resulted because editor
conventions vary from program to program. These difficulties have
dampened somewhat our initial enthusiasm for the TENEX system.

Nonetheless, TENEX provides some features which are crucial
to a comfortable network environment. The standard support programs included with this system facilitate both the sending of
messages to other users (either at the same site or at other sites
on the ARPA network) and the transfer of data and programs from
site to site on that network; also the ability to "link" two or
more terminals allows users to communicate easily and immediately.
Both the linking and message facility have been found to be
invaluable aids in inter-group communications and in such problems
as interactive program debugging. When two terminals are linked,
their output streams are merged, thus allowing each terminal to
display everything typed at the other terminal. Since only the
output stream is affected under these circumstances, it is still
possible for each terminal to be used to provide input to separate
programs, in addition to being used in a conversational mode.

Resource allocation. As noted above, the computational
resources of the SUMEX-AIM facility are apportioned by the AIM
advisory group and SUMEX management. Some extensions to the basic
TENEX system have been made to reflect this apportioning in the
actual use of the facility. Basically, it was recognized that
users of the facility are members of groups working on specific
projects, and it is among these projects that the facility is
apportioned. Disk space and cpu cycles are now distributed among
groups instead of among individual users. For example, a user may
exceed his individual disk allocation somewhat without any ill
effect, so long as the total allocation of his group remains
within the limits. Similarly, a Reserve Allocation Scheduler has
been added to TENEX which tries to match the administrative cycle
distribution over a ninety second time frame. Thus a particular
group cannot dominate the machine if a lot of its members are
logged in at one time.

It is typical for usage of a facility to peak through the
middle hours of the day. Indeed, one of the advantages of having
users from around the country is the spreading of the load caused
by the difference in time zones. Even so, the facility could
offer better service if more people would shift their main usage
hours toward either end of the day. To encourage "soft-schedul-
ing" within groups on the system, SUMEX-AIM publishes a weekly
plot of diurnal loading. This plot shows the total number of jobs
on the system as well as the number of LISP jobs, since these jobs
seem to make the biggest demands of system resources. The result

has been an increased awareness by users of system loading and a
noticeable increase in the number of users at all hours of the
night and early morning.

Protection and system security. Protection for a computer
system covers a range of ideas. It means the ability to maintain
secrecy - for example, to guarantee the privacy of patient records.
It also guarantees integrity by assuring that programs and data
are not modified by an unauthorized party.
   Questions of protection generally become more interesting and
complex as more sharing is involved. Consider the example of a
proprietory program which generates layouts given a user's circuit
data. The program owner demands assurance that he will be paid
whenever his program is used and that copies of the program cannot
be made. The user wants guarantees that his data sets cannot be
destroyed or copied for a competitor. Yet the user must have
access to the program and the program must have access to the data.
Unable to support such complicated examples of protection, SUMEX-
AIM assumes that sharing takes place between friendly users. This
is not to imply that issues of protection and sharing have not
appeared. For example, in an effort to improve the human engineering of programs for public use, the capability of recording a
session has been built into several of the programs. Studied by
the programs' designers to pinpoint confusing aspects of programs,
these recordings serve to improve program design. Since the issue
of violation of privacy has been raised, some of these programs
now request permission to record a session beford doing so. At
this time, any guarantee of privacy must be provided by the program designer because TENEX itself does not have the ability to
render protection.
   The general design for systems offering "state of the art"
protection involves a tolerance for failure; that is, if a potential offender succeeds in breaking through some of the defenses,
he still does not place the entire computer system at his mercy.
Encrypting of data files provides an additional line of defense.
This method is used by at least two calendar or appointment programs on the computer. At this time, however, there are no
general encrypting facilities available and users must do this for
themselves as needed.
   TENEX provides the usual keyword protection at login time and
a measure of file protection. Owners of a file may assign a protection number which specifies some combination of READ, WRITE,
EXECUTE, or APPEND access to a file for owners, members of a
group, or other users. This level of protection is basically
enough to prevent accidents and most mischief. System programmer's
around the country are aware of a number of TENEX bugs which permit this access to be violated. One user of our system found a
way to place himself in a mode where he could modify any file on
the system. To date, we have no examples of such activity
actually having a deleterious effect on SUMEX-AIM.

To make the use of SUMEX-AIM programs easy, on a trial basis for prospective users, a 'guest' account system has been established. Since this makes logging into SUMEX-AIM so easy, it has invited some misuse by people using this account to play computer games. A proposed extension to the system now being implemented is a special "guest EXEC" which would extend the protection of the TENEX monitor by allowing guest accounts access to only a more restricted set of programs.

File backup. In order to assure the user maximum protection against loss of valuable work, SUMEX operates a multi-level file backup system. In addition to a routine file backup system, there are facilities to enable the user to selectively archive his or her disk files. By issuing a simple command to the TENEX executive, the user can transmit a message to the operator to copy specified files to magnetic tape. Each such file is copied to two magnetic tapes within 24 hours of issuing the archive command. File retrieval is affected by a similar process. The user also has the alternative option of being able to lodge files in a special backup directory. Files are held in this directory until the next exclusive file dump (see below) at which time they are deleted. In this way the user can remove files from his directory at his own choosing knowing they will be archived by the exclusive dump.

On a system level, an effort is made to maintain file backups such that the maximum possible loss, in the event of a crash fatal to the file system, would amount to no more that one day's work. Once each day all files that have been read or written within the last 48 hours are dumped onto magnetic tape. Files that exist for 48 hours are thus held on two separate tapes. The rotation period for files dumped in this way is 60 days. Once each week a full file dump is made to separate disk storage. Each such dump is kept for two weeks at which time it is replaced by a new file dump. Each month there is a full system dump from disk to magnetic tape. Files can be recovered from the system backup by sending a message to the operator specifying the file name(s) and when the file was last read or written (if such information is available).

Excessive demand for production programs. One of the concepts behind the creation of a shared resource is elimination of the problems which arise when large, complex computer programs are exported. Since, in theory, exportability is no longer a problem, there is greater latitude in choice of a language in which program development can take place. In the case of some of the DENDRAL programs, it was thought that program development should take place in INTERLISP, a language that lends itself well to the artificial intelligence nature of these programs, but does not lead to particularly efficient run-time code.

In order to ascertain the usefulness of these programs and to determine what areas remain in need of work, chemist collaborators are being sought. As these users increase in number and begin to use the programs more frequently, it is almost certain that the inherent slowness of the predominately LISP code will affect the whole system as well as handicap the efficient use of the DENDRAL programs. Additionally, some of the chemist-users who are finding the programs most useful and who are most enthusiastic about their potential use, are persons working in industry. Although, in one sense, this interest from industry could be interpreted as an indication of the "real-world" usefulness of the programs, it came as rather a surprise to both SUMEX and DENDRAL personnel.

The fact that SUMEX-AIM is funded by NIH as a national resource prohibits the facility from providing a service, at taxpayer's expense, to a private industry. Although there is precedent for a site funded via government grant to charge a fee for service, such an arrangement leads to highly complicated bookkeeping, and is contrary to the essential purpose of SUMEX-AIM; to be a research-oriented rather than a service-oriented facility. This leaves the industrial users in the position of being more than willing to pay for the use of the programs, but of having no mechanism whereby they can be charged. Furthermore, the fact that the programs are coded in LISP for a highly specialized environment, almost guarantees the impossibility of export, except to an almost identical computer system.

An intermediate solution that will help to solve the problem of industrial users on SUMEX and will help to alleviate the system loading resulting from heavy usage of LISP coded production programs, is to mount CONGEN on a closely related computer which is operated on a fee for service basis. However, in order to make this program available at a reasonable fee, it has become evident that it will be necessary to recode the LISP sections of the program into a more efficient and easily exportable language.

Research-oriented Problems

Community mindedness. Those involved in computer science research at SUMEX face a general problem which is absent or greatly lessened at non-network sites; the problem of community mindedness. The network provides a large and varied set of other researchers and users who have an interest in their work. Although the network-TENEX combination provides new forms of communication with these remote parties, the traditional means of fully describing the use and structure of a complex program, a detailed person-to-person discussion, is not convenient. Comprehensive documentation gains importance in such a situation, and within the DENDRAL project a great deal of time has been needed in the development of program descriptions which are adequate for a diverse audience. Also, in both DENDRAL and MYCIN, effort has been and is being directed toward "human engineering" in program design; to provide the user with commands which assist him in using the programs, in understanding the logic by which the programs reach certain decisions and in communicating questions or comments on the programs' operation to those responsible for development. Such "housekeeping" tasks can often be neglected, yet are quite important in smoothing interaction with the community.

Choice of programming language. High level programming languages which are designed for ease of program development are frequently poor as production-level languages. This is because developmental languages free the researcher from a raft of programming details, thus allowing him to concentrate upon the central logical issues of the problem, but the automatic handling of these details is seldom optimal. Also, because such languages tend to be specialized for certain computers and operating systems, the exportation of programs can be a serious problem. One solution to these problems is the recoding of research-level programs into more efficient language when fast and exportable versions are needed.

Networking greatly eases the problem of exportability, but can also aggrevate the problem of efficiency. As mentioned in the previous section, DENDRAL programs, which are undergoing constant development, found a substantial number of production-level users. Because of the inefficiencies of INTERLISP (a 50- to 100-fold improvement in running time is not uncommon when an INTERLISP program is translated into FORTRAN), this use adversely impacted the entire system. Because the DENDRAL programs are quite large and complex, their translation into other languages is impractically tedious. A partial solution to this problem is provided by the TENEX operating system, which allows some interface between programs written in different languages. With such intercommunication, time-consuming segments of an INTERLISP program which are not undergoing active development can be reprogrammed in another more efficient language. The developmental parts of the program are left in INTERLISP, where modifications can easily be made and tested. The CONGEN program uses three languages; INTERLISP, FORTRAN and SAIL (26). The SAIL segment was added when a new feature, whose implementation was fairly straightforward, was included in CONGEN. Since then, the SAIL portion gradually has been taking over some of the more time-consuming tasks. This method allows a balance in the tradeoff between ease of program development and efficiency of the final program.

Accumulation of expert knowledge in knowledge-based programs. Just as statistics-based programs need to worry about accumulation of large data bases, knowledge-based programs need to worry about the accumulation of large amounts of expertise. The performance of these programs is tied directly to the amount of knowledge they have about the task domain -- in a phrase, knowledge is power. Therefore, one of the goals of artificial intelligence research is to build systems that not only perform as well as an expert but that also can accumulate knowledge from several experts.

Simple accretion of knowledge is possible only when the "facts", or inference rules, that are being added to the program are entirely separate from one another. It is unreasonable to expect a body of knowledge to be so well organized that the facts or rules do not overlap. (If is were so well organized, it is unlikely that an artificial intelligence program would be the best encoding of the problem solver.) One way of dealing with the overlap is to examine the new rules on an individual basis, as they are added to the system in order to remove the overlap. This was the strategy for developing the early DENDRAL programs. However, it is very inefficient and becomes increasingly more difficult as the body of knowledge grows.

The problem of removing conflicts, or potential conflicts, from overlapping rules becomes more acute when more that one expert adds new rules to the knowledge base. Of course, the advantages of allowing several experts to "teach" the system are enormous -- not only is the program's breadth of knowledge potentially greater than that of a single expert, but the rules are more apt to be refined when looked at by several experts. On the other hand, one can expect not only a greater volume of new rules but a higher percentage of conflicts when several experts are adding rules.

Having a computer program that can accumulate knowledge presupposes having an organization of the program and its knowledge base that allows accumulation. If the knowledge is built into the program as sequences of low-level program statements -- as often happens -- then changing the program becomes impossible. Thus, current artificial intelligence research stresses the importance of separating problem-solving knowledge from the control structure of the program that uses that knowledge.

Another problem, at a political rather than a programming level, becomes apparent with one accumulation process: how does the program distinguish an expert from a novice? In the MYCIN program we have circumvented the problem by having the program ask the current user for a keyword that would identify him as an expert. It is then a bureaucratic decision as to which users are given that keyword. There is nothing subtle in this solution, and one can imagine far better schemes for accomplishing the same thing. The point here is that not every user should have the privilege of changing rules that experts have added to the system, and that some safeguards must be implemented.

"Human nature" barriers to SUMEX use

Countering disbelief. There is sometimes a tendency among those unfamiliar with the capabilities and limitations of computers and computer programs to express disbelief. This is not disbelief in the sense of worrying that the programs have errors and produce erroneous results. Indeed, the fact that a problem is being done by a computer seems to generate some faith that it might be right, or at least significantly reduces questions about correctness. The disbelief is that programs, which are designed to model, or to emulate, human problem solving will not be capable of useful performance. This, of course, is the classic argument against artificial intelligence - we think in mysterious ways and have such a complex brain that a computer program must be inferior. In some cases, authors of artificial intelligence programs have brought such criticism upon themselves by not stressing limitations, or by making extravagant claims.

In the DENDRAL project, we have tried to counter this type of disbelief in a number of ways. We have tried to stress that our programs are designed to assist, not replace chemists. We have always discussed limitations to give a reasonable perspective on capabilities vs. limitations of a program. Most importantly however, we have focused on those aspects of problems which are amenable to systematic analysis, i.e., those problems which can be done manually, but only with difficulty and with the consumption of a great deal of time which a chemist could better spend on more productive pursuits. Examples of this would include the application of PLANNER to mixtures where all fragmentations may have to be considered as possible fragments of every molecular ion, the systematic analysis by INTSUM of possible fragmentation processes, the consideration by MOLION of all plausible possibilities, and the structure generation capabilities of CONGEN.

We have also tried to reduce chemists' disbelief by blurring the "outsider-insider" distinction, in particular by having trained chemists work on the programs and make them useful to themselves first. Further, when "outside" chemists are first introduced to the programs, the introduction is done by another chemist who has already thought through and can readily explain many of the chemistry-related problems.

The ultimate way to counter disbelief, however, is to illustrate high levels of performance. If a potential user is aware of the goals (intent) of a program and its limitations, a few examples of results which would be extremely difficult to obtain without the program are very convincing.

The "security" of a local facility. Networking is still a relatively new concept to many people, and there is a resistance to departing from the "traditional" modes of computing. There is a sense of security in having a local computing facility with knowledgeable consultants within walking distance, and in having "hard" forms of input (e.g., boxes of computer cards) and output (e.g., voluminous listings). These props are difficult to simulate over a network connection - in most cases a user's interaction with the remote site takes place exclusively through a computer terminal - yet the quality of service can match or exceed that of a local facility; programs and large data sets can be entered and stored on secondary storage as can large output files; all types of program and data editing can be done with interactive editing programs; programs can be written in an interactive mode so that small amounts of control information can be input and key results output in "real time" over the terminal; and as noted in a previous section, consultation can be significantly more productive providing that the remote operating systems supports the appropriate types of communication possibilities.

There can, of course, be no denying that there are problems in learning to use a distant computer system, be it for program development or for the use of certain programs. Whether or not overcoming these problems to gain access to the special resources which are available, is worth the effort, is a question answerable only by the individuals involved. Fortunately, there will always be those persons who have a pressing problem in need of solution and who are willing to try a new approach; regardless of whether or not they have had prior network experience.

IV. THE SUMEX-AIM FACILITY

The SUMEX-AIM computer facility consists of a Digital Equipment Corporation model KI-10 central processor operating under the TENEX time sharing monitor. It is located at Stanford University Medical Center, Stanford, California.

The system has 256K words (36 bit) of high speed memory; 1.6 million words of swapping storage; 70 million words of disk storage; two 9-track, 800 bpi industry tape units; one dual DEC-tape unit; a line printer; and communications network interfaces providing user terminal access via both TYMNET and ARPANET.

Software support has evolved, and will continue to evolve, based on user research goals and requirements. Major user languages currently include INTERLISP, SAIL, FORTRAN-10, BLISS-10, BASIC and MACRO-10. Major software packages available include OMNIGRAPH, for graphics support of multiple terminal types, and MLAB, for mathematical modelling.

The SUMEX-AIM computer generally is left with no operator in attendance; thereby helping to eliminate some overhead, but also creating some problems. Users who wish to run jobs requiring tapes must make arrangements to mount their own tapes. Likewise, obtaining listings from the line printer can be somewhat difficult since there is no regular schedule for distribution of this output. The solution to these two problems has been to make keys to the machine room available at strategic locations, convenient to all groups of local users. This experiment in basic "resource sharing" has not resulted in any of the major problems one might expect from having a fairly large group of people with hands-on access to a computer.

Acknowledgements

The SUMEX-AIM facility is supported by the National Institutes of Health (grant number RR 00785-02), as is the DENDRAL project (grant number RR 00612-05A1). We wish to thank Robert Engelmore, Mark Stefik, Janice Aikens and Peter Friedland for their contributions to this report and Tom Rindfleisch and William White for valuable discussions.

Literature Cited

(1) Gordon, R. M., Datamation(1975), 21, (2), 127.
(2) "World List of Crystallographic Computer Programs", Second Edition, D. P. Shoemaker, Ed., Bronder-Offset, Rotterdam, 1966.
(3) Professor Joshua Lederberg, Principal Investigator.
(4) Lederberg, J., Sutherland, G. L., Buchanan, B. G., Feigenbaum, E. A., Robertson, A. V., Duffield, A. M. and Djerassi, C., J. Amer. Chem. Soc.(1969), 91, 2973.
(5) Duffield, A. M., Robertson, A. V., Djerassi, C., Buchanan, B. G., Sutherland, G. L., Feigenbaum, E. A. and Lederberg, J., J. Amer. Chem. Soc.(1969), 91, 2977.
(6) Buchanan, B. G., Duffield, A. M. and Robertson, A. V., "Mass Spectrometry: Techniques and Applications," G. W. A. Milne, Ed., p. 121, John Wiley and Sons, New York, 1971.
(7) Dromey, R. G., unpublished results, preprint available on request, Dept. of Computer Science, Serra House, Stanford University, Stanford, Calif. 94305.
(8) Biller, J. E. and Biemann, K., Anal. Lett.(1974), 7, 515.
(9) Several libraries of mass spectral data are available in various forms. The Aldermaston Data Centre (Mass Spectrometry Data Center, AWRE, Aldermaston, Reading RG7 4PR, England) can provide information on the availability of such libraries.
(10) Hertz, H. S., Hites, R. A. and Biemann, K., Anal. Chem. (1971), 43, 681.
(11) Dromey, R. G., Buchanan, B. G., Smith, D. H., Lederberg, J. and Djerassi, C., J. Org. Chem.(1975), 40, 770.
(12) Smith, D. H., Buchanan, B. G., Engelmore, R. S., Duffield, A. M., Yeo, A., Feigenbaum, E. A., Lederberg, J. and Djerassi, C., J. Amer. Chem. Soc.(1972), 94, 5962.
(13) Smith, D. H., Buchanan, B. G., Engelmore, R. S., Aldercreutz, H. and Djerassi, C., J. Amer. Chem. Soc.(1973), 95, 6078.
(14) Smith, D. H. and Carhart, R. E., Abstracts, 169th Meeting of the American Chemical Society, Philadelphia, April 6-11, 1975.
(15) Carhart, R. E., Smith, D. H., Brown, H. and Djerassi, C., J. Amer. Chem. Soc., submitted for publication.
(16) Masinter, L. M., Sridharan, N. S., Lederberg, J. and Smith, D. H., J. Amer. Chem. Soc.(1974). 96, 7702.
(17) Masinter, L. M., Sridharan, N. S., Carhart, R. E., and Smith, D. H., J. Amer. Chem. Soc.(1974), 96, 7714.
(18) Brown, H., SIAM Journal on Computing, submitted for

publication.
(19) Wipke, W. T. and Dyott, T. M., J. Amer. Chem. Soc.(1974), 96, 4825.
(20) Smith, D. H., Buchanan, B. G., White, W. C., Feigenbaum, E. A., Lederberg, J. and Djerassi, C., Tetrahedron(1973), 29, 3117.
(21) Buchanan, B. G., to appear in the Proceedings of the NATO Advanced Study Institute on Computer Oriented Learning Processes, 1974, Bonas, France.
(22) Carhart, R. E., Smith, D. H. and Brown, H., J. Chem. Inf. Comp. Sci., in press (May, 1975).
(23) Smith, D. H., Anal. Chem., in press (May, 1975).
(24) Teitelman, W., "INTERLISP Reference Manual," Xerox Corp. (Palo Alto Research Center), Palo Alto, Calif., 1974.
(25) Bobrow, D. G., Burchfiel, J. D. and Tomlinson, R. S., Commun. ACM(1972), 15, (3), 135.
(26) VanLehn, K. A., "SAIL User Manual," Stanford Artificial Intelligence Laboratory, Stanford, Calif., 1973.

14

Networks for Research Sharing

JAMES C. EMERY

Executive Director, Planning Council on Computing in Education and Research, Interuniversity Communications Council, Inc. (EDUCOM),
P. O. Box 364, Princeton, N. J. 08540

The importance of the computer continues to grow in terms of the number of applications, number of users served, level of resources expended, and the increasingly crucial role it plays in organizations and intellectual activities. An organization or professional worker denied access to the best computing resources will suffer increased penalties as these trends continue.

The explosive advances in computer technology and the diversity of applications make it exceedingly difficult for any one person or organization to keep up with the computer field. Current budget pressures often severely limit the resources available for exploring ways in which the computer can increase an organization's efficiency or effectiveness. Faced with this situation, many organizations are looking to resource sharing as a means to take advantage of the full range of the computer's power while also limiting their expenditures in trying to keep up with the march of technology.

Benefits of Sharing

The traditional argument for sharing computing power is the substantial economy of scale exhibited by computers. Grosch hypothesized in the late 1940's that raw capacity increased in proportion to the square of cost, and this relationship has held reasonably well over the past 25 years. Very large compute-bound

---

This reasearch was supported in part by the Office of Naval Research under Contract #N0014-67-A-0216-0007.

jobs will continue to benefit from this phenomenon, and so the sharing of a large processor among many users will continue to offer attractive economies.

Other hardware economies are also possible through sharing. Sharing allows a user with large but intermittent demands to avoid the high cost of maintaining his own, largely idle, capacity. Sharing many demands of this sort can result in a much higher capacity utilization than each user could achieve on his own. This same argument applies to intermittent peak loads: by maintaining only the capacity necessary to meet the minimum "base load" demands and then using a shared computer to satisfy peak demands, capacity utilization can be increased substantially. If jobs are freely transportable among multiple computers connected through a network, greater utilization can be achieved by shifting jobs from an overloaded computer to one having idle capacity.

Although these economies should not be ignored, they are becoming of much less importance than they were formerly. Hardware costs, and the direct costs of operating a facility, continue to decline as a fraction of the overall cost of computing. Other costs -- data collection, storage, communications, remote terminals, and design and maintenance of the system -- have become the dominant factors in most current systems. The cost of raw CPU power has become almost irrelevant in many cases, and advances in electronics will continue this trend.

Does this mean that the sharing of computing resources has become irrelevant? Not at all. The sharing of software and databases will continue to offer very substantial benefits. Indeed, the very technical advances that have given us such low-cost processors now make it economically feasible to operate a widespread network for sharing programs and data. The shift in costs from hardware to other components will increase the relative importance of this form of sharing.

Such sharing spreads the cost of developing, operating, and maintaining the programs and databases. It also avoids problems of inconsistencies and reconciliation that inevitably arise if results from separate programs or databases must be compared. In some cases a program or database may offer unique capabilities that can only be accessed through sharing.

Ways of Sharing

Resources can be shared through a variety of

means. The traditional way of sharing software and data is by replicating the resource at the user's own computer center. A statistical package or a database consisting of corporate financial data, for example, might be recorded on magnetic tape and transported to each computer center serving users who want to access the shared resource. This has proved extremely useful, but it certainly has some serious limitations. Incompatible hardware or systems software often make transportability quite difficult -- even among machines of the same manufacturer and model number. The maintenance of duplicate programs is usually a non-trivial problem. Providing training and other services to remote users is difficult and expensive -- and therefore usually not done very well. With few exceptions, successful examples of transported resources have been achieved by an organization that had a strong incentive to promote transportability and was willing to provide the necessary effort and resources to make it happen.

A network that provides remote users access to a shared resource overcomes some (but by no means all) of the limitations of program and database transportability. It also permits the sharing of hardware resources as well as programs or data. If a user wants to access a program or database, the computation normally takes place at the site where the program or database is maintained, but in some cases a user might prefer to transmit the resource to his own local computer and perform the computation there.

Network operation may be either centralized or decentralized. A centralized network may consist of one or more computers* governed as a single entity. Hardware and software are usually standardized, and all users must conform to the standard. In contrast, a decentralized network consists of multiple computers, each of which is governed independently.

Centralized management of the network encourages standardization and efficiency. A decision can be made on the basis of its effect on the network as a whole, rather than on only one part of it. For example, the number and configuration of computers in the network, and the allocation of tasks among them, can be determined on the basis of the best balance between cost, reliability, response time, and the like.

---

* A single computer serving remote terminals has been labeled a _star_ network, although there is some question whether this should be regarded as a network.

What may appear advantageous in principle, however, may not always prove to be attractive in practice. Optimization of a complex network is an enormously complex task, and the centralized decision process may not be able to cope with it. Managers of a large, centralized network may become less responsive to user needs than to their own (perhaps limited) criteria of efficiency. The organizations served by the network may prefer to participate in its management instead of turning over this task to a centralized group.

These disadvantages have largely restricted the centralized approach to networks established by a single organization to sell external services or to serve its own subunits; very few independent organizations have banded together to form a centralized network to serve their combined needs. Universities, in particular, have seldom chosen this path except through the application of external incentives or pressure (by a State legislature, for example).

In the summer of 1974, eighteen universities (soon to be twenty) agreed to form a Planning Council on Computing in Education and Research. Organized within the corporate structure of EDUCOM, the Council has established as its primary goal the exploration and development of ways to improve the efficiency and effectiveness of computing within higher education. A network will very likely play an important role in meeting this objective. Any such network will undoubtedly consist of decentralized computer centers operated by independent universities and colleges.

Costs of Sharing

Sharing, though very attractive in a number of circumstances, does not come for free. When hardware is shared among multiple users, some means must exist for allocating resources among jobs. At the detailed level, the computer's operating system must allocate CPU time, primary and auxiliary storage, input/output facilities, and the like; at a higher level, sharing entails extra effort to determine the combined hardware and software needs of users, establish pricing policies to allocate costs, define job priorities, and similar matters concerned with the joint use of a shared resource.

Sharing of a computing resource can be expanded -- with the aim to increase the economies of scale -- by attracting a broader group of users. This necessarily increases the generality of the system, often at a considerable cost. For example, a computer that combines scientific computing with data processing might well result in useful economies of scale; the generality and complexity of the combined system will certainly be greater than if each group of users were served by a separate machine. Generality in software -- for example, a statistical analysis package that offers a large number of options in order to serve a wide range of users -- fosters sharing. It does this, though, at a cost in design time, operating efficiency, and complexity.

Rather than increasing a system's generality in order to attract a wider class of users, one might alternatively require the users to adapt to a common standard as a way of increasing resource sharing. This avoids some costs of generality, but might impose on users an extra cost of changing to the new standard, living with a system that does not meet their needs as well as a more tailored resource, or simply doing without a desirable service. In any case, the standardization is not achieved without cost.

Sharing can also be expanded by increasing the size of the geographical area served instead of increasing the intensity of use within a given area. This allows the use of a relatively specialized resource -- and thereby gaining the economies of specialization (or avoiding the diseconomies of generalization) -- while at the same time permitting a large enough load to exploit some economies of scale. Thus, a nationwide or worldwide group of users having a common interest might well justify developing tailored hardware or software to serve that specialized market. The cost of such sharing is the cost of communications over a widely dispersed area, as well as the added cost of providing marketing and other support services to a group of remote users.

Administration of a network adds further to the cost of resource sharing. Contractual arrangements must be made with users and suppliers of service. Authorization procedures, which define who has access to what resources, must be established and operated. Information about available resources must be collected, maintained, and disseminated.

A potential cost of sharing is the loss of autonomy and control brought about by joint use of a resource. Scheduling problems and the need to adapt to a common standard often attend such sharing. Scheduling delays and compromise standards that serve no one group very well could in some cases impose serious costs. These costs can mount if the network is administered centrally with a narrow view of "efficiency" and an insufficient motivation to remain responsive to user needs (which can easily happen, for example, if the network serves a captured audience having no alternative source of computing services).

Characteristics of a "Facilitating" Network for Higher Education

The design of a network must recognize the tradeoffs among the costs and benefits of resource sharing. Each situation may call for a unique set of decisions regarding the users to be served, services offered, appropriate degree of centralization and standardization, and administrative arrangements. These decisions are always subject to change in light of more detailed design work or operating experience.

The computing community within higher education -- students, faculty members, researchers, and administrators -- has a number of special needs; a network designed to serve this population must recognize this fact. Each institution views itself as autonomous and insists on remaining so. Individuals within each institution are themselves often independent of effective central control. Users range from the naive to the super-sophisticated. Applications cover an extremely wide spectrum of academic disciplines, computing techniques, and level of resources required.

In view of these special requirements, one can outline desirable characteristics of a network for higher education:

- The network will preserve the autonomy of each institution that supplies or purchases services on the network.
- Hardware, programs, and databases will be distributed geographically and administered decentrally.
- The network will not have a monopoly as a buyer or seller of services -- i.e., each buyer and seller will be free to deal with others outside of the network (although voluntary exclusive contracts would not be precluded).
- Entry and exit from the network will be open to all institutions.
- A market mechanism will be the primary means of allocating resources of the network.
- A wide variety of specialized and proprietary services will be offered over the network.

- Network administrative functions -- authorization of users, billing and reporting, dissemination of information about available resources, and similar services aimed at facilitating relations between buyers and sellers -- will be available from a central network organization. Buyers and sellers will avail themselves of these facilitating services only to the extent that it serves their interests to do so.
- The costs of operating the network and providing computing resources must be borne by the users of the network.

Use of a Facilitating Network

The availability of a national facilitating network will open up important new opportunities for colleges and universities. Not all institutions will use the network in the same way. Some institutions might choose to become totally dependent on the network for computing services, doing away with their own local computers (except for terminals with varying degrees of "intelligence"). Others may not buy services to any significant extent, but rather use the network to market their services to external users. And, indeed, other institutions will decide to ignore the network, choosing to be neither a buyer nor seller of external computing resources.

The typical large institution will probably be both a buyer and a seller of service. It will no doubt serve the bulk of its users with its own local computer. The emphasis is likely to be placed on meeting the assured "base load" requirements, and then meeting peak overloads from the network. Relatively simple, low cost applications -- such as routine batch processing and small interactive programs -- will be handled locally. The institution can gear up to satisfy these needs in a very cost-effective manner -- for example, with a purchased machine configured for efficient batch processing and kept fully loaded over a long lifetime, or by a minicomputer dedicated to providing relatively limited time-sharing capabilities (only programs written in BASIC, say). Most computer users on a campus could be satisfied within the local base load capacity, while relatively few (probably fewer than 20 percent) would have to resort to the network to obtain more specialized services.

Despite the fact that few users would require network services, the availability of such services would permit substantial economies. Traditionally, an institution's computer center has felt obliged to offer a wide range of services, because users had no other source. If, however, the network alternative exists, the choice of which service to offer locally can be resolved largely on technical and economic grounds. Specialized services that are rarely required typically absorb a disportionate share of costs at most computer centers. If these low-volume, high-cost applications can be served through the network, substantial reductions can be effected in hardware, programming, and data collection costs. The supplier of a specialized service can combine the demands from many institutions in order to spread the high cost of providing the service.

Not all sophisticated and specialized services will be obtained from the network; most institutions will choose to offer services in which they have a special interest or unique capabilities. Network sales of these services to external users might often offset a university's purchases of other services from other institutions. Thus, a university could limit its <u>net</u> external purchases while at the same time permitting its users to take advantage of a vastly expanded array of services available from the network.

Sharing over a national network would offer to all institutions the best computing resources available throughout the country. Even a modest-sized college or university might obtain access to large-scale computers or hardware designed for specialized applications. It could take advantage of an extensive library of programs for performing standard analyses or specialized applications in such diverse areas as financial planning models for university administrators and computer-assisted instruction material. Its faculty and students could retrieve information from a wide range of databases that deal with such matters as econometric and demographic data, bibliographic references, and physical characteristics of materials.

Not even the largest and most heavily endowed universities could hope to provide such a wide range of resources except through sharing. But smaller institutions perhaps stand to gain even more than large ones. The sharing of resources permits a small college to retain its identity and individuality while at the same time permitting it to offer the same computational resources available at a large institution. Without access to shared resources, it would be at a severe disadvantage in attracting the best faculty and students.

Some regional and statewide networks have already been established as a means of gaining the advantages of sharing. Although they have certainly expanded the services available to users, they are still limited in their capabilities. Such networks thus become prime candidates for linking to a "distributed" network that combines local, regional, and national resource sharing.

Problems to Overcome

As compelling as the argument is for such sharing, the fact is that relatively little of it currently takes place among educational institutions. To some extent, this stems from the creative nature of educational and research institutions that encourages students and faculty to discover things on their own. It is easy to overemphasize such resistance to sharing, however; far more serious have been the limitations of earlier technology and administrative impediments to effective sharing.

The transportability of programs and data has always proved more difficult than one might suppose. As a result, it is often easier to recreate a program or database than it is to find it at some other location and transport it to a local facility or use it remotely.

Because of the limited sharing that currently takes place, little incentive exists for faculty members or their institutions to exert the extra effort required to make their resources more transportable. In order for a program to be used elsewhere, it must be carefully tested and debugged, exhaustively documented, and faithfully maintained and extended. The education of potential users and even direct personal assistance are often necessary. Proprietary firms have been relatively successful in providing services to remote users, but only because they have had the economic incentives to do so; members of the academic community typically have had neither financial nor professional incentives to share computational resources.

It is quite clear that extensive sharing will not take place until a better mechanism exists to foster it. One of the important technical requirements is a communications network that links educational institutions. The network must permit ready and timely access to all of the available resources at each of the connected computer centers. A prospective user must also have access to information about what resources are available and their prices. If the user runs into difficulties in using a remote resource, he should be able to obtain assistance and instructions.

The technology for developing such a network already largely exists. Computer operating systems have developed to a point that permits widespread sharing among many users and programs. Commercial "packet switching" networks and other recent advances in communications technology will provide highly reliable and relatively low-cost data transmission. Furthermore, the cost of communications in an advanced network can be made largely independent of distance; as a result, geographical separation will cease to inhibit sharing to the extent that it has in the past.

The principal problems that remain are chiefly administrative and motivational. In order for the network to function with any hope of success, a means must exist to coordinate network activities, provide communications services, establish prices, disseminate information about available resources, simplify access procedures, provide incentives to suppliers of resources, control access to the network, provide security and privacy, and handle the billing for services rendered.

The network will rely primarily on a market mechanism to allocate resources and encourage suppliers to remain responsive to user needs. Conditions within the network -- many buyers and sellers and up-to-date information about the services available and their prices (with perhaps a quality rating as well) -- approximate the economists' definition of a perfectly competitive market with its attendant efficiency in allocating resources.

Pricing obviously plays a critical role in such a network. Suppliers would have an incentive to set prices that encourage efficient use of their resources. For example, price differentials should motivate some users to shift their demands from peak to off-peak periods in order to impose a more level load on the system. The pricing structure should be flexible enough to permit buyers and sellers to satisfy their needs within a level of risk they are willing to assume. For example, long term contracts should be permitted to reduce uncertainty regarding the availability of service and the prices that will be charged. On the other hand, "spot" sales at any price above marginal cost should also be permitted to encourage the use of excess capacity (although this opens up some very difficult issues concerning full versus marginal pricing and cost accounting regulations of the Federal government).

Conclusions

Given the independent nature of educational institutions, it is almost certain that a monolithic, centralized computing network would be neither feasible nor desirable. Connection to the network, and the use or supplying of network resources, will remain a matter governed by decentralized decisions. Each institution will be guided by its own self-interest. But in pursuing its own interests, an institution will have an incentive to become a supplier of services for which it enjoys a comparative advantage and to purchase services that it can obtain more cheaply, or with higher quality, than duplicating them locally. Thus, the invisible hand of Adam Smith, rather than the heavy hand of a centralized bureaucrat, will provide the primary means of fostering the exchange of services.

A "facilitating network" provides all of the advantages of decentralized operation, while also creating vast opportunities that would otherwise not exist. Not only would it greatly broaden the resources available to users of computers and databases, but it could also open exciting possibilities for persons who have so far not benefited much from computer technology. As the cost of terminals and communications goes down, it is not at all inconceivable that virtually every faculty member and student will have access to a network through his own terminal.

A widespread network of this sort could be used to communicate among scholars as well as with computers. Even the most confirmed non-quantitative person could take advantage of some of the network services -- for example, an "electronic mailbox" that allows him to communicate rapidly and inexpensively with remote collaborators. Each person would thus have access to both computer-based and human resources throughout the country and, indeed, the world. The usefulness of the so-called "invisible college," the informal person-to-person network that has proven so effective in the exchange of information among scholars, would be augmented immeasurably.

The development of a mature network will come about through an extended and cautious evolutionary process. At each stage in its development the system must prove its worth by sustaining itself financially. Growth in usage will be gradual as institutions accommodate themselves to new opportunities and make appropriate adjustments in their local computer centers. There seems to be little doubt, however, that a widespread network will evolve. The arguments for it are compelling, and we can see no insuperable technical, economic, or administrative problems that would prevent its development.

INDEX

A

Acetylcholine	10
Adams–Bashforth	36
Adams–Bashforth predictor–corrector integration	39
Administrative data processing (ADP)	146
ADP, administrative data processing	146
Air Research Automatic Computer, OARAC Office of	155
Allosteric effects	45
Analysis	
conformational	28
electrical system	46
Fourier	46
mechanical system	46
Angles, bond lengths and	30
Applications	
models of	41
of network conferencing, geologic	53
specific biomolecular	43
ARPA	53, 58
ARPANET	9, 42, 111, 156
Array processor, Floating Point Systems'	36
ASD Computer Center, background on the	154
Assembled testing in a large network, computer	129
ASSIGN	120
Assignment, core memory	104
Associative processing	24
Astrophysics	22
Atomic motion	22
Atomic positions	27

B

Barrel–roll discriminator	33, 36
Bashforth, Adams–	36
Batch-processing, open-shop	93
BCM	87
Beam scattering, molecular	30
BINARY	75
BISON	163
BISONMC	163, 166
Bond lengths and angles	30
Bond strengths	30
Born–Oppenheimer	28
Box, twinkle	26
Brookhaven National Laboratory	4
BUILDER	3
Bus structure, memory	102

C

CALCOMP	92
Calculation, maximum likelihood	24
Calculational Techniques	23
Calculations, quantum mechanical	28
CAMAC	37, 40, 42, 48
Cambridge Crystal Data File	6
Cancer Research, Institute for	4
Capacity measurements, heat	100
Card records, job control	170
Carnegie-Mellon Multi-Mini Processor	25
Catalysis, enzyme	128
CATC	150
CD	30
CDC–CTBERNET	9
Central, network	68
Centralized/localized computing	111
Channel interface	111
Characteristics, network programming	73
Charge distributions	30
Chemical	
applications of interactive computing	197
kinetics	46
physics	20
reactions and evolution of molecular systems, dynamics	41
Chemistry data bank	136, 137
Chemistry, polymer	21
CHESS	120
CIMS, Courant Institute of Mathematical Science	11
City University of New York (CUNY)	11
CLEANUP	199
CLEANUP, spectrum output by	201
C.mnp	25
Code(s)	
hollerith	13
installation and maintenance of	162
reentrant	104
COGEODATA	63
Collaborative research community	192
Collisions, short-range	23
Communication(s)	
and control, user	89
dyadic	56, 58
interprocessor	37
terminal	103
Complexity, particle	23
Computation, long-term	93
Computation, molecular dynamics	31

Computational system 31
Computer(s)
  assembled testing in a large network 129
  configuration 88
  graphics, remote terminal 9
  identification and interpretation of
    unknown mass spectra 183
  network, hierarchical 36, 42
  network of real-time 67
  network on college chemistry de-
    partments, the impact of 142
  paper tape based 73
  resource sharing, history of 154
  utility for interactive instrument
    control 85
Computing
  centralized/localized 111
  CRYSET: A Network for
    Crystallographic 1
  distributed 78
Concurrent user operations 91
CONDUIT 140, 142, 147, 148
Conferencing, geologic application of
  network 53
Conferencing system 53
Configuration
  computer 88
  hardware 88
  U–wide computer network 119
Conformation, protein 21
Conformational analysis 28
CONGEN 204, 207
Connections, diagram of interactive
  terminal 165
Control
  card records, job 170
  computer utility for interactive
    instrument 85
  points, SOCRATES modules and 133
  programs, custom designed 89
Core memory assignment 104
Coulomb 30
Courant Institute of Mathematical
  Sciences (CIMS) 11
CRT 26
CRYSET: A Network for Crystallo-
  graphic Computing 1
CRYSNET Vector General 3
Crystal Data File, Cambridge 6
Crystallographic Computing,
  CRYSNET: A Network for 1
Crystallography 1
  project, protein 196
  x-ray 9
CUNY, City University of New York 11
Current interfaced instruments 95
Current on-line experiments 94
Custom designed control programs 89
CYBER 70/76 .... 1, 2
CYBERNET 140

D

Data
  bank, chemistry 136
  protein 6
  usage of the chemistry 137
  file, Cambridge crystal 6
  flow 103
  processing (ADP), administrative 146
  transfer 88
DECOMP 120
DENDRAL 187, 192, 197, 215
  programs 198
  project 196
Depth, well 30
Description tables, program 104
Design, prosthetic device 46
Detection, nuclear particle 99
Detection systems, multi-particle 106
D–glucaro–3–lactone, space-filling
  model of 5
DIALOG project 195
Differential equations, solution of
  coupled 24
Differential transmission scheme 122
Diffraction, x-ray 30
Diffractometer, neutron 96, 106
Digital and hybrid graphics systems 26
Digital voltmeter (DVM) readings 99
Discriminator 34
Discriminator, barrel–roll 33, 36
Disk sector, logical 90
Dispersion forces, London 30
Distance, internuclear 30
Distributed computing 78
Distribution of molecular parameters,
  statistics 43
Distributions, charge 30
DVM readings, digital voltmeter 99
Dyadic communication 56, 58
Dynamics 18
  chemical reactions and evolution of
    molecular systems 41
  computation, molecular 31
  molecular 9, 20, 22
  particle 22

E

EBCIDIC-ASCII 104
EDUCOM 140
Effects, allosteric 45
Effects, scale of 23
Electrical system analysis 46
Electron-density map 6
Electronic spectroscopy 30
ELLIPS 2
Empirical and semi-empirical energy
  functions 28
ENDOR and ESR spectroscopy 98

INDEX

Energy functions, empirical and semi-empirical ... 28
Energy surface, potential ... 20
Entry, intelligence remote job ... 11
Enzyme catalysis ... 128
Enzyme kinetics ... 46
Equations, solution of coupled differential ... 24
Equilibrium ... 22
Equilibrium geometry ... 30
Ergodic theorem ... 43
Evolution of molecular systems, dynamics—chemical reactions and ... 41
Examples, paper tape program ... 74
Execution, non-resident program ... 90
Expansion, incremental ... 94
Experiment(s)
  interfaces ... 103
  current on-line ... 94
  with the U.S.G.S. ... 58

F

Facilities, graphics ... 92
Feely, Touchy ... 26, 33, 44
Floating point array processors ... 36
Flow, data ... 103
Force(s) ... 18
  functions ... 20, 27
  functions, interatomic potential or interatomic ... 29
  interatomic ... 20
  London dispersion ... 30
  theoretical ... 29
FORTRAN IV-H ... 87, 93
FORUM ... 53, 54
  system at U.S.G.S. ... 59, 60
  system, current experiments with the ... 53
Fourier
  analysis ... 46
  transformations ... 85
  transform methods ... 23
Functions
  empirical and semi-empirical energy ... 28
  force ... 20, 27
  interatomic potential (or force) ... 20, 29
  protein ... 45
  trigonometric ... 21

G

Geologic applications of network conferencing ... 53
Geometry, equilibrium ... 30
Graphics
  facilities ... 92
  molecular ... 2
  remote terminal computer ... 9
  systems, digital and hybrid ... 26

GRASP ... 59
GROPE-1 ... 27

H

Hardware ... 11, 14
  configuration ... 88
  network ... 69
  processors ... 26
Heat capacity measurements ... 100
Hemoglobin, stereoview of lamprey ... 7
Hierarchical computer network ... 36, 42
Hierarchical minicomputer support ... 108
Hollerith code ... 13
Hunter Chemistry Computer Laboratory ... 15
Hybrid graphics systems, digital and ... 26

I

Identification and interpretation of unknown mass spectra, computer ... 183
Illiac IV ... 24
Incremental expansion ... 94
INFONET ... 58
Infrared and Raman spectroscopy, vibrational ... 30
Infrared spectroscopy, UV, visible and ... 100
Installation and maintenance of codes ... 162
Institute for Cancer Research ... 4
Instructions, general user ... 62
Instrument control, computer utility for interactive ... 85
Instruments, current interfaced ... 95
Integration, Adams–Bashforth predictor–corrector ... 39
Interaction complexity ... 23
Interaction, scale of ... 23
Interactive computing, chemical applications of ... 197
Interactive terminal connections, diagram of ... 165
Interatomic
  forces ... 20
  potential (or force) functions ... 20, 29
Interface
  channel ... 111
  molecular manipulation—touch ... 33
  molecular portrayal—visual ... 33
  touch ... 26
  visual ... 26
Interfaced instruments, current ... 95
Interfaces, experiment ... 103
Interfaces, visual and touch ... 36
Internal rotation, barriers to ... 30
International networking ... 63
Internuclear distance ... 30
Interpretation of unknown mass spectra, computer identification and ... 183

Interpretive and Retrieval System,
  Self-Training 186
Interprocessor communication 37
INTSUM 207
Isomerization 21
ITG module, SOCRATES 140

J

JCCRGEN example 172, 175
JCCR's 170
JEOL 121
JEOLJMS 121
JMS–D100 121
Job
  control card records 170
  entry, intelligence remote 11
  entry, RJE (remote) 141
Johns Hopkins University 4

K

Kinetics
  chemical 46
  enzymes 46
  stopped-flow 98

L

Laboratory methodology 112
Langrangian 18
LAOCOON III 120
Law, Newton's Second 27
LDS 90
Lengths and angles, bond 30
Level–flow process, liquid 127
Lincoln Wand system 26
Line Notation, Wiswesser 186
Liquid level–flow process 127
LISTING 75
Lockheed Differential Equation
  Processor 24
Logical disk sector 90
London dispersion forces 30
Long-term computation 93
L–phenylalanine cation, drawing of ... 5
L–phenylalanine hydrochloride 5

M

Macronetwork 118
Magnetic resonance facility, repetitive
  scanning 122
Magnetic resonance spectroscopy 9
Maintenance of codes, installation and 162
Manipulation—touch interface,
  molecular 33
MANIPL 3
MANIPL, output of program 7
Map, electron-density 6
Mass memory operating system
  support 77

Mass spectra, computer identification
  and interpretation of unknown 183
Master integrals and SCF codes,
  development of 156
Match factor 186
Maximum likelihood calculation 24
Measurements, heat capacity 100
Mechanical system analysis 46
Mechanics
  multiprocessor molecular 17
  statistical 22
  stellar and plasma 22
Medical Foundation of Buffalo 4
Medical Information Systems
  Laboratory 195
Membrane transport 45
Memory
  assignment, core 104
  bus structure 102
  operating system support, mass 77
  protection 102
Meta–4 .... 39, 40
Methods, Fourier transform 23
Methodology, laboratory 112
MF 186
MHBOOT 121
Microde 33
Microphone, strip 26
Microwave spectroscopy, rotational ... 30
Minicomputer(s)
  interfacing support system (MISS) 109
  network of real-time 67
  support, hierarchical 108
Mininet, UMR 119
MIOP 89, 90
MIPS 25
MISL project 195
MISS, minicomputer interfacing
  support system 109
Modes of application 41
Molecular
  beam scattering 30
  dynamics 9, 20–22
  computation 31
  graphics 2
  manipulation—touch interface 33
  mechanics, multiprocessor 17
  parameters, statistics—distribution
    of 43
  portrayal—visual interface 33
  statics 20
  structure, statics 41
  systems, dynamics—chemical reactions and evolution of 41
MOLION 202
Momentum 22
Motion, atomic 22
Multi-mini Processor, Carnegie–
  Mellon 25
Multi-particle detection systems 106

INDEX

Multiprocessor molecular mechanics ... 17
Multiprocessor network ... 33
MYCIN project ... 196

N

Network
  central ... 68
  conferencing, geologic applications
    of ... 53
  configuration ... 72
  configuration, U-wide computer ... 119
  hardware ... 69
  hierarchical computer ... 36, 42
  multiprocessor ... 33
  of real-time mini computers ... 67
  programming characteristics ... 73
Networking at UMR, computer ... 118
Networking, international ... 63
Neutron diffractometer ... 96, 106
NEWTON ... 25, 30, 33, 36–43
  macroscopic machine ... 17
Newton's Second Law ... 27
NMR ... 30
  spectrometer ... 106
Non-resident program execution ... 90
Notation, Wiswesser line ... 186
Nuclear magnetic resonance
  spectroscopy, pulsed ... 95
Nuclear particle detection ... 99
Nuclease, staphylococcal ... 6

O

On-line experiments, current ... 94
Open-shop batch-processing ... 93
Operations, concurrent user ... 91
Operations, vector ... 21
Optical processing ... 24
ORD ... 30
ORFFE-3 ... 97
ORFLS-3 ... 97
ORTEP ... 10
Output ... 164

P

PAINT ... 165
Paper tape based computers ... 73
Paper tape program examples ... 74
Parallel Element Processing
  Ensemble (PEPE) ... 25
Parameters, statistics—distribution of
  molecular ... 43
PBM ... 184, 188
Particle(s)
  complexity ... 23
  detection, nuclear ... 99
  dynamics of ... 22
Pattern recognition ... 24
Pen, sonic ... 26
PEPE ... 25

Pharmacology–toxicology ... 46
Physics, chemical ... 20
Physics, plasma ... 22
PLANET ... 53
PLANET-1 ... 56
PLANNER ... 204, 207
Plasma ... 23
  mechanics, stellar and ... 22
  particle calculations ... 7
  physics ... 22
Plate reading, spectroscopic ... 99
PLATO ... 141
Plotter, Versatec ... 4
Polarizabilities ... 30
Polymer chemistry ... 21
Polymers, biological ... 19
Portrayal—visual interface, molecular ... 33
Positions, atomic ... 27
Potential energy surface ... 20
Potential functions, interatomic ... 20
Predictor–corrector integration,
  Adams–Bashforth ... 39
PRJCTN ... 2, 4
Probability based matching ... 184
Process, liquid level-flow ... 127
Processing
  associative ... 24
  open-shop batch- ... 93
  optical ... 24
  signal ... 24
Processor(s)
  array ... 36
  Carnegie–Mellon Multi-Mini ... 25
  floating point array ... 36
  Floating Point Systems' array ... 36
  hardware ... 26
  Lockheed differential equation ... 24
  supervisory ... 33
  terminal interface ... 156
Program(s)
  custom designed control ... 89
  DENDRAL ... 198
  description tables ... 104
  execution, non-resident ... 90
Programming characteristics, network ... 73
Project
  DENDRAL ... 196
  DIALOG ... 195
  MISL ... 195
  MYCIN ... 196
  protein crystallography ... 196
Proline ring ... 7
Prosthetic device design ... 46
Protection, memory ... 102
Protein(s) ... 20
  conformation ... 21
  crystallography project ... 196
  data bank ... 6
  function ... 45
  structure ... 45

Pulse radiolysis ... 97
Pulsed nuclear magnetic resonance spectroscopy ... 95

Q

Quantum mechanical calculations ... 28
Quantum mechanics ... 9

R

RAD ... 90, 93
Radiolysis, pulse ... 97
Raman spectroscopy, vibrational (infrared and) ... 30
Reactions and evolution of molecular systems, dynamics—chemical ... 41
Reading, spectroscopic plate ... 99
Readings, digital voltmeter (DMV) ... 99
Real-time mini computers, network of ... 67
Real-time response ... 89
Records, job control card ... 170
Recognition, pattern ... 24
Reentrant code ... 104
Regional computer center (RCC) ... 142
RCC, regional computer center ... 142
Remote
 job entry, RJE ... 141
 terminal choices ... 162
 terminal computer graphics ... 9
Research community, collaborative ... 192
Resource sharing, history of computer ... 154
Resource sharing, networks for ... 219
Response, real-time ... 89
Retrieval System, Self-Training Interpretive and ... 186
Retrieval system, SOCRATES ... 130
RJE (remote job entry) ... 141
Rotation, barriers to internal ... 30
Rotation (microwave) spectroscopy ... 30
RULEGEN ... 207

S

Sample transcript ... 61, 64
Scale of effects ... 22
Scale of interaction ... 23
Scanners ... 26
Scanning magnetic resonance facility, repetitive ... 122
Scattering, molecular beam ... 30
SCF codes, development of master integrals and ... 156
Scheduling ... 105
Scheme, differential transmission ... 122
Second Law, Newton's ... 27
Sector, logical disk ... 90
Self-Training Interpretive and Retrieval System ... 186
Semi-empirical energy functions, empirical and ... 28
Services, system ... 88

Sharing, history of computer resource ... 154
Sharing, networks for resource ... 219
Short-range collisions ... 24
Signal processing ... 24
Signal, ultrasonic ... 26
SOCRATES
 ITG module ... 140
 modules and control points ... 133
 retrieval system ... 130
 system ... 129
Software ... 11, 14
Solution of coupled differential equations ... 24
Sonic pen ... 26
Spectrometer, NMR ... 106
Spectroscopic plate reading ... 99
Spectroscopy ... 30, 31
 electronic ... 30
 magnetic resonance ... 9
 pulsed nuclear magnetic resonance ... 95
 rotational (microwave) ... 30
 UV, visible and infrared ... 100
Stanford University Medical Experimental computer ... 192
Staphylococcal nuclease ... 6, 7
Statics ... 18
 molecular ... 20, 21
 molecular structure ... 41
Statistical mechanics ... 22
Statistical thermodynamics ... 9
Statistics—distribution of molecular parameters ... 43
Stellar and plasma mechanics ... 22
STIPL ... 3, 5
STIRS ... 186, 187, 188
Stopped-flow kinetics ... 98
Storage ... 167
Strengths, bond ... 30
Strip microphone ... 26
Structure
 dynamics–function ... 47
 memory bus ... 102
 protein ... 45
 statics, molecular ... 41
SUMEX-AIM ... 192, 193, 216
 access to ... 194
 community, interactions in ... 194
Support, hierarchical minicomputer ... 108
Support, mass memory operating system ... 77
Supervisory processor ... 33
Surface, potential energy ... 20
SYMBOL ... 87
System(s) ... 26
 analysis, electrical ... 46
 analysis, mechanical ... 46
 computational ... 31
 conferencing ... 53
 digital and hybrid graphics ... 26

INDEX

dynamics—chemical reactions and
    evolution of molecular 41
Lincoln Wand 26
multi-particle detection 106
services 88
support, mass memory operating ... 77

T

Tables, program description 104
Tape based computers, paper 73
Tape program examples, paper 74
Techniques, calculational 23
Tektronix 92
Tektronix 4010 storage tube display
    terminal 14
Teleconferencing 53
Teleprocessing 9
Terminal
    communications 103
connections, diagram of interactive 165
interface processor 156
Testing in a large network, computer
    assembled 129
Texas A & M University 4
Theorem, Ergodic 43
Theoretical, forces 29
Thermochemistry 30
Thermodynamics, statistical 9
Thymidine-3',5-diphosphate,
    stereoview of 7
TI980A 121, 125, 128
Time-sharing 93
TIP 156
Touch interface 26
Touch interface, molecular
    manipulation— 33
Touch interfaces, visual and 36
Touchy Feely 26
Touchy Feely I 33, 44
Touchy Feely II 33
Touchy Twisty 26
Touchy Twisty I 33
Toxicology, pharmacology 46
Transcript, sample 61, 64
Transfer, data 88
Transformation, Fourier 85
Transmission scheme, differential .. 122

Transport, membrane 45
Trigonometric functions 21
Twisty, Touchy 26, 33
Twinkle Box 26
TYMNET 183
TYMSHARE 140

U

Ultrasonic signal 26
UMR, computer networking at 118
UMR mininet 119
User communication and control .. 89
User instructions, general 62
U.S.G.S., experiments with the 58
U.S.G.S., FORUM system at 59, 60
Utility for interactive instrument
    control, computer 85
UV, visible, and infrared spectroscopy 100
U-wide computer network
    configuration 119

V

Van der Waals 30
Vector operations 21
Vector general, CRYSNET 3
Versatec plotter 4
Vibrational (infrared and Raman)
    spectroscopy 30
Vickers Wand 26
Viscosity 22
Visible, and infrared spectroscopy, UV 100
Visual interface 26
Visual interface, molecular
    portrayal— 33
Visual and touch interfaces 36
Voltmeter (DVM) readings 99

W

Wand system, Lincoln 26
Wand, Vickers 26
Well depth 30
WLN 186

X

X-ray crystallography 9
X-ray diffraction 30

QD
39.3
E46
C64

APR 19 1978